MANUAL SCAVENGING IN INDIA

About the Author

B.N. Srivastava, M.A. (Economics) is an eminent social activist, administrator and planner. He served the Government of India and held several prestigious positions including in the Ministry of Home Affairs, Ministry of Welfare, etc. As a planner, he formulated strategies for rapid development of weaker sections. He made valuable contributions in the implementations of several developmental programmes meant for weaker sections. At present, Shri Srivastava is Hony. Chairman (Administration), and Member-Secretary, Non-official Commission on Scavenging Sulabh International Social Service Organisation.

All along, Shri Srivastava has had a concern for the downtrodden, particularly scavengers. While in Government he conceived and chalked out several new schemes for their training and rehabilitation. He has successfully implemented several projects. Side by side he, participated in many symposia, conferences and workshops within and outside the country. He is a member in the roster of international experts in low cost sanitation by the UNCHS (Habitat). He is a resource person to several governmental and non-governmental organisations.

A prolific writer, Shri Srivastava has to his credit books and research articles published in journals from India and abroad. His current interests are Human Rights and Social change and development.

MANUAL SCAVENGING IN INDIA

A Disgrace to the Country

B.N. SRIVASTAVA
Former Director
Ministry of Home Affairs/Welfare
Govt. of India

Published for
SULABH INTERNATIONAL SOCIAL SERVICE ORGANISATION, NEW DELHI

by

CONCEPT PUBLISHING COMPANY, NEW DELHI-110059

ISBN 81-7022-639-2

First Published 1997

Published and Printed by
Ashok Kumar Mittal
Concept Publishing Company
A/15-16, Commercial Block, Mohan Garden,
NEW DELHI - 110059 (India)

Phones : 5554042, 5504042
Fax No. : 91-11-5598898

CONTENTS

PREFACE

The sweeper or scavenger has many synonyms, known by different names in different parts of the country. As a scavenger or rather as a 'sweeper up of dust' the *Chuhra* is termed *'Khakrob'*. As a domestic servant he is ironically styled *Mehtar* or 'Chieftain' which is a honorific title of various classes such as *'Bhatiyara', 'Qasai',* etc, as a worker in leather he is called a *'Dhed'* (lit. crow) in Gujarat, as a weaver he is styled, *'Megh'* at least in Rajasthan and, as an executioner he is known as *'Jallad'*. Further as a tanner, the *Chuhra* is called a *'Khatik'* in Punjab and as a breeder of swine he is known as a *'Hali'*.

The name *'Mehtar'* or Prince was commonly applied to the servants of the Emperor Humayun. Another title for them was *'Halalkhor,* "one who eats what is lawful, one whose earnings are legitimate". According to W.Crooke,this euphemistic title is said to have been introduced by the Emperor Akbar. The other title is 'Lalbegis' who lived on the outskirts of villages and towns and were not allowed to enter the locality by day. It is said that during the reign of Aurangzeb, a corpse was found lying in the streets of Delhi and it appeared to Muslims to be a Hindu corpse and to Hindus to be a Muslim one. The Lalbegis were at last ordered to remove it but as they went to carry it away, it became converted into a heap of flowers. This they claimed to be their *Pir* Lalbeg, who had come to help them out of their degraded position. The Emperor was so struck with wonder at the change that he allowed the Lalbegis to live in towns and villages and to carry on their profession by day as well as by night.

My concern to the cause of these unfortunate people dates back to the late sixties when I was deputed by the Government of India to study the living conditions of sweepers and scavengers in the municipalities of Andhra Pradesh. During my tours of Nalgonda, Khammam, Eluru, Rajamundry, Machhlipatnam and Vyzaiwada I found that the quarters allotted to the sweepers and scavengers were in the most neglected condition. The location was filthiest spot—very often near and around a public latrine, pail-depot, a dirty drain or a dumping ground and far away from their place of work. It was also seen that there were no arrangements for ventilation, water, school, roads etc. No toilets or bathing places were provided to them. My concern about the pitiable condition of these people was highlighted by the Parliamentary Committee on the Welfare of Scheduled Castes and Scheduled Tribes in its report to the Parliament in 1970.

In March 1981, a meeting was held in the room of Shri Bhisham Narayan Singh, the then Union Minister of Works and Housing to discuss the proposal of the Government of Bihar for taking up a scheme for the liberation of scavengers in some selected towns of the State. Besides Dr. Bindeshwar Pathak of Sulabh International who took the initiative for organising this meeting, it was attended by the representatives of the Government of Bihar and the Union Ministries of Home Affairs and Works & Housing. As a Director in the Ministry of Home Affairs, I also attended this meeting. Since there was no scheme in the Ministry of Home Affairs where this proposal could be accommodated, it was proposed to include it under the Centrally Sponsored Scheme of 'Implementation of PCR Act' as one of the measures for the removal of untouchability since the scavengers were the victims of the worst form of untouchability because of their unclean occupation (under Section 15-A of the PCR Act, 1955). A lot of hectic exercise was undertaken to get the scheme approved by the Planning Commission and the Finance with the twin objectives of conversion of existing dry latrines into pour-flush sanitary latrines and rehabilitation of liberated scavengers in alternative employment/ occupation. The scheme was finally approved and Central assistance was released to the State Government in respect of only two towns during that year due to limitation of funds. Subsequently the scheme was taken up in other States as well. The scheme worked very well and during the period 1981-1991, about one million dry latrines were converted into sanitary ones, 15,000 liberated scavengers were rehabilitated in alternative employment and 40 towns were declared scavenging-free. This was definitely a good achievement.

In this small book, an attempt has been made to present an accurate and fairly complete account of the scavenging community. The sweepers and scavengers in India *are the stigmatized people whose stigma is derived from birth-ascribed group membership and is shared throughout the group* (Berreman, 1967). Such people exist in many societies but the most prominent examples are the Blacks of USA, the Eta of Japan and the Dalits of India. The book is dedicated to Padma Bhushan Dr. Bindeshwar Pathak for the splendid work done by him in this field. He is most passionately concerned about the problems of transition which are facing the simple Balmiki people. It is not merely a matter of emotion for him. He is a man of action, mobilising the people, the administration and the political system to achieve that objective. He has an insatiable desire to make India scavenging-free.

I am indebted to Dr. A.K. Lal of the A.N. Sinha Institute of Social Studies, Patna, who inspired me to write this book on the occasion of the Silver Jubilee Year of Sulabh Movement which also coincided with the 125th birth anniversary of Mahatma Gandhi. I am very much thankful to him. I have no words to express my gratitude to Dr. Bindeshwar Pathak who took keen interest in this work and provided valuable guidance from time to time. Lastly, I must thank my better-half, Mrs. Asha Srivastava who was a source of constant inspiration to me in completing this project.

New Delhi B.N. SRIVASTAVA

1

INTRODUCTION

The Indian social structure is a unique social system which is not found in any other part of the world. In other parts of Asia, Europe, America and Africa etc. although there is some system of social stratification, but those are quite different in nature and based on economic, social and racial set-up. In India, the caste system is a very complex phenomenon. Each caste stands for a way of life that is to some extent distinctive, but at the same time the caste of a region forms part of a single social framework. Caste is ubiquitous and this has resulted in an ideology tolerant of diversity.

All old vedic literary accounts confirm that there were four major castes/groups in the society in the past, namely, Brahmins, Kshatriyas, Vaishyas, and Shudras. The fact that all these four categories have been discussed in all the literature available during a longer period and ascribing the fact that the origin of all the four groups are of divine source. These classes or order were regularly referred to in later literature as *Varnas*. *Varnas* basically meant colour and it was to define the contrasting features of different groups of people particularly in relation to fair Arya with dark *Dasa*. In *Rig Veda* there have been several references of the three former groups and the reference of the fourth group (Shudra) occurs only once. In the post vedic-period, the reference regarding Shudras have been made and they were shown as less privileged. The occupation of the first two *Varnas* are clearly stated to be priesthood and administrative and military duties respectively. But the duties of the Vaishya and Shudra are not very clear. The village headman was usually a Vaishya and Shudras were servants. The *Varna* categories do not however exhaust the various occupations practised in vedic India like chariot-builder, gold-smith, barber, physician, leather-worker, potter, merchants and others. It is not clear how they formed part of the *varna* framework.

The post-vedic period witnessed the growth and consolidation of the power of Brahmins. They being at the top of the *chaturvarna,* defined the duties and rights of each caste and its place in the social hierarchy. This is evident from the epics like *Ramayana* and *Mahabharata*. Even in the *Bhagwat-Gita* the caste system has been justified on the basis of the ideas of *guna, karma* and *dharma.*

The main features of the Indian caste system as categorised by Prof. Srinivas* (1969) are:

(i) Hierarchy (born in a particular caste and keep separate identity by using caste names).
(ii) Endogamy and Hypergamy (marry within the caste and not outside).
(iii) Occupational association (caste is identified with a particular profession/occupation).
(iv) Restriction on food, drink and smoking.
(v) Distinction in custom, dress and speech.
(vi) Pollution (by sight or touch).
(vii) Ritual and other privileges and disabilities.
(viii) Caste organisation (caste council and panchayats etc.).
(ix) Caste mobility (caste habitations).

The essence of caste is characterised by the presence of hereditary groups in a hierarchy with Brahmin at the top and Dalit at the bottom. There is a caste in Tamil Nadu known as Paraiyan who were associated in the past, with playing the drum on certain occasions. In fact drumming is associated with untouchable castes in many parts of the country. The other occupations of the Paraiyan were weaving and farm labour. In social rank they were by no means the lowest for there were several other castes like Chakkilian etc. which were considered inferior to the Paraiyan. But they were undoubtedly low enough in the hierarchy to be treated as untouchables. Like other untouchable castes they were considered polluted and unfit for residence inside the village alongside the clean castes. Max Weber adopted Pariah as a generic term for stigmatized, impure, downgraded people found in many societies. He identified Pariah groups more or less with guest residents and defined them as "underprivileged social groups who, though their occupations were economically indispensable for their host societies were nevertheless treated as 'impure' and consequently ritually avoided".

These Pariah groups are treated as 'untouchables', a category

---

* *India: Social Structure,* Prof. M.N. Srinivas.

that includes a number of castes, all collectively treated as impure and outsiders. They live on the fringe of the society, often excluded from social contact with its members though their services are accepted as highly valuable and often necessary for the survival of the society.

The Indian Pariahs are called by different names such as 'untouchables', 'Harijans' (a glorified term, coined by Narsimha Mehta and adopted and popularised by Mahatma Gandhi), 'Exterior Castes' (as called by J.H. Hutton), 'Depressed Classes', 'Outcastes'. and 'Dalits'. Officially these Pariah groups are comprised under Scheduled Castes'. The term 'Scheduled Castes' appeared for the first time in the Government of India Act 1935, the British Government issued the Government of India (Scheduled Castes) Order 1936, specifying certain castes, races and tribes as Scheduled Castes. The test applied for inclusion in the list was the extreme social, educational and economic backwardness arising out of the traditional practice of untouchability.

As said earlier, one of the features of the Indian caste system is that each caste has been traditionally associated with a particular occupation. The castes which consider themselves superior to others enjoy a wider range of choices in occupations compared to other castes. The Kshatriyas or Brahmins could engage themselves in any occupation like agriculture, industry, education, trade or business besides their caste occupations. Vaishyas too had a variety of choices although most of them stick to petty trading and money-lending. But in the case of other castes specialisation was more marked where work required special aptitude as well as skill. They had no choice but to stick to their hereditary occupations because of the training they got from their parents and grand parents, investments in tools and equipment and a reasonably secured customary rights and other incomes.

Unclean Occupation

The Dalits, too had traditional occupations but with a distinct difference. By and large their occupations were least desirable and defiling. The two main categories of Dalit occupations were cleaning and leather processing. Cleaning or sweeping involved sweeping of streets, drains and sewers; removal of human and animal excreta; raising of scavenger pigs etc. Since latrines in most cities were not provided with flush, the sweeper had to carry buckets of human faeces on his or her head. Various castes* of Valmikis who are considered to

* Annexure I.1

be lowest in the caste hierarchy, did this work. Leather work included handling of dead animals, removing their skin and curing them, and making of leather goods such as shoes, chappals etc. All of these tasks were considered defiling by the so called clean castes. By assigning these tasks to people who were already polluted the society spared the normal people. The reasoning was then easily reversed and the Dalit's pollution was attributed to the unclean nature of their work. The Dalit jobs are not only undesirable, polluting and tedious, but they are also low paid. To eat the left over from another persons' plate is considered highly degrading. It is common practice among sweepers to collect every day one or two *chapatis* from the houses they serve. Although this is a part of their customary wage, it is considered almost to begging. This practice of collecting *chapatis* is gradually disappearing from big cities like Delhi, Bombay etc. but it is still very much prevalent in other small towns.

Earlier Literature and Origin of Scavenging

The earlier literature does not mention any caste like the Bhangis, remover of night-soil, among the traditional occupations nor do they mention of any caste exclusively relegated to this profession. Even to this day scavenging as a profession does not exist in most of the rural areas.

The institutionalisation of 'sweeping and scavenging' as a profession seems to be of recent origin. There is a reference in the Kautalya's *Arthashastra* that during Maurya period defecation in an open space in the towns was prohibited but there is no mention about the disposal of night-soil by scavengers. The old scriptures, however, throw some light on the system of scavenging. Dr. Bindeshwar Pathak (1991) writes in his *'Road to Freedom'* that "according to the contents of sacred scriptures and other literatures, scavenging, especially the disposal of night-soil by a particular caste or castes of Indian society, has been in existence since the beginning of civilization. One of the fifteen duties for slaves enumerated in the *Naradiya Samhita,* was to dispose of human excreta. In *Vajasaneyi Samhita* the Chandals and Paulkasa have been referred to as slaves for the disposal of night-soil". It is alleged that only after the advent of the Muslims in India that 'sweeping and scavenging' seems to have taken the form of a formal profession. It is said that the system of bucket privies was designed and constructed by Muslims for their women in *pardah.* Those who were made captives, were forced to clean latrines, bucket privies and throw off the night-soil at distant places. These captives when freed, were not accepted by the society and they formed a separate caste and continued the work of scavenging. During the British period with the setting up of army cantonments and municipalities, a large number of people were perhaps, required to do these services on a regular basis.

Functional Castes

The names of scavenging castes suggest that they are a functional community recruited from many different racial and social groups. It is very likely that people belonging to the lowest strata were compelled to take on this profession especially in the urban areas due to economic necessity. The great variations in the physical features of the different sections of this community, but also the varied recruitment from higher castes of "broken men"—outcastes, destitutes etc. show that the members of various castes have joined this profession at different points of time obviously due to economic compulsions as this occupation was regarded as unpleasant by the economically better-off sections. This only proves that the members of different castes and communities have taken up to this vocation mainly due to economic constraints.

The sweepers will freely admit into their community, outsiders of whatever caste for some serious offence with no chance of reconciliation. Various methods are employed for the initiation of a neophyte which vary from region to region. Owing to this re-inforcement from higher castes, the sweepers are generally taller and stronger as well as lighter in colour than the average Hindu or Muslim.

Stephen Fuchs* (1981) has pointed out that "the endogamous sub-sections of this lowest of all low-castes, are not without a certain social gradation. Within the scavenging caste some sections are superior to others, and the social status is determined according to the origin of the section, claiming a former higher rank, or according to the type of work a particular section is doing. The lowest place is generally occupied by those sections which carry night-soil. The highest section have given up scavenging entirely, and have adopted other occupations, such as leather work, weaving, or farming. Thus we see even those castes that are entirely outside the community of castes, are very much influenced by the Hindu caste spirit and grade themselves according to the same principle generally valid among the true Hindu castes. Thus even, the so called outcastes or 'Exterior

* *At the Bottom of Indian Society*—Stephen Fuchs.

castes' have established a hierarchy of their own, though it might be completely ignored by the higher castes who may treat all the scavenging castes alike—as impure, polluting outsiders with whom they want to have no social contact though they accept their services and indeed, are dependent on their services."

He further writes that "among those lowest scavenging sections which remove night-soil there is still a distinction; those who serve in private houses consider themselves higher than those who clean public latrines. It should not be forgotten that this type of work has to be carried out in urban environment and in villages where the women are not permitted to leave the house, even where it is for a call of nature".

"Each of these sub-sections has its own rules and regulations as to what work is permitted and when the line is to be drawn. Some do not touch night-soil, but do all the other scavenging work in the villages, including the removal of dead animals and refuse. Other sections work as porters or play drums and other musical instruments for the villagers; again others have become sweepers of the roads, farm labourers, watchmen, cane labourers, even domestic servants, tanners and leather workers or weavers".

"In the west, where the caste of sweepers seems to have the lowest status, the sweeper will handle a corpse, kill a stray dog and act as hangman. Further east, the scavenger caste refuses these services, and they are carried out by other castes, by the Doms, or Mangs, for instance".

In order to escape the curse of untouchability, a large number of sweepers mostly in northern India, have abandoned the Hindu religion and embraced Islam, Christianity or Sikh religions. But this change of religion has not brought any significant change in their occupation or social status though these religions reject the caste system and profess equality of all men. This has, however, infused a feeling of greater solidarity amongst them towards their own community than towards their religion. For example, a Hindu Chuhra will readily marry a Chuhra girl belonging to Muslim or Christian religion and vice versa, while they are reluctant to marry a mate of the same faith, but outside their caste.

All Scavengers are not Scheduled Castes

Most of the people think that the sweepers and scavengers belong only to the Scheduled Castes. According to the Constitution (Scheduled Castes) order, 1950 as amended from time to time "no person who profess a religion different from the Hindu, the Sikh or Buddhist religion shall be deemed to be a member of a Scheduled Caste". This means that the person professing religion other than Hindu, Sikh or Buddhist like Christianity, Islam or tribal cannot be Scheduled Castes. It would be perhaps revealing that according to 1961 Census* only 48.2 per cent of the total number of persons engaged in sweeping, scavenging and allied occupations belonged to the category of Scheduled Castes. It was found that the proportion of Scheduled Castes engaged in sweeping and scavenging to the total population** engaged in this occupation varies from 5.58% in Jammu and Kashmir to 90.43% in Himachal Pradesh. A survey@ conducted by the Committee on Improvement of Living and Working Conditions of Sweepers and Scavengers headed by Shri I.P.D. Salappa on the sweepers or Pourakarmiks in Karnataka working in the various municipalities in the State, shows that besides the Scheduled Castes, there are Muslims, Christians, Lingayats, Kuruba and Mudaliars engaged in this profession. This only proves that members of different castes and communities have taken up to this vocation mainly due to economic compulsions.

Scavengers in the List of Backward Classes

However, the members of those occupational groups which were not included in the list of Scheduled Castes, have now been included in the Central list of Backward Classes in accordance with the Supreme Court Judgement of November 1992 in the Indira Sawhney case (popularly known as Mandal Judgement) and now entitled for all the benefits which are admissible to the Backward Classes by the Central and State Governments.

---

* Census of India 1961, 'Social Mobility among Sweepers of India.

** Annexure I.2

@ Usha Rao, N.J: *Deprived Castes in India* (1981).

2

HISTORICAL BACKGROUND

Social Mobility for Higher Status

The British administration was attracted by the institution of castes in India partly for political reasons and partly for the scientific interest evoked by its ubiquity and uniqueness. On the basis of census data, Risley, the Commissioner for 1901 Census, classified castes into seven main categories according to their social standing and ranked the *jatis* in the local hierarchy and *varna* affiliation of each. The ranking of *jatis* and castes by the census created an unprecedented situation. Whatever their *de facto* status, most of the communities at the lower rung of the caste ladder felt that it was a good opportunity for social climbing by laying claims to higher status and registering a higher ranking in the census documents to have an official stamp, indicative of their higher social origin. Thus was set in motion a process of social mobility whereby a caste, claimed a higher social status by sanskritising its religious beliefs and rituals, if necessary. A number of caste associations were formed and overnight honorific caste names were adopted, showing descent from Brahmins or Rajputs etc.

The scavenging* castes which were known by different names in different States like Bhangi, Balmiki, Chuhra, Mehtar, Mazhabi, Lal Begi, Halalkhor etc. in northern India; Hari, Hadi, Hela, Dom and Sanei etc. in eastern India; Mukhiyar, Thoti, Chachati, Pakay, Relli etc. in Southern India; and Mehtar, Bhangias, Halalkhor, Ghasi, Olgana, Zadmalli, Barvashia, Metariya, Jamphoda and Mela etc. in Western and Central India; also made an effort to get united and have a common name. In 1911 census some of them started returning as Adi Dharmi, Adi Dravida, Adi Karnataka and Adi Andhra. Hutton, in 1931 estimated the population of various scavenging castes to the

* Annexure I.1

extent of 20 lakhs* in the undivided India. After independence, the census in 1961** estimated their population to be 8.2 lakhs. The number of scavengers who were engaged in the removal of night-soil was 3,86,725 which can be projected as 6.18 lakhs for 1981. The Task Force constituted by the Planning Commission had estimated*** the number of scavengers in respect of Scheduled Castes for the year 1989 as 4.0 lakhs.

Scavenging Castes of Northern India

The sweepers and scavengers of Northern India go under various names. The most common name is that of 'Mehtar' which is really a title meaning 'prince' or leader. To quote an example, the head of the ruling family of Chitral State (now in Pakistan) is called 'Mehtar of Chitral'. Another name for this caste is Bhangi. The term 'Bhangi' is derived from the Sanskrit word *bhangi* meaning hemp; it seems to be an allusion to their drinking habits. Others derive the word from *bhangi,* meaning broken. This would, however, only fit those scavengers who split bamboos and plait mats and baskets. In Punjab the scavengers are known as Churas. This name probably comes from *Chura jharna,* to sweep scraps. Another name common in Punjab and Uttar Pradesh is Balmiki or Lal Begi. These are the names of two great saints, the first being a Hindu and the other a Muslim. The other names are Hela, Hari, Hadi, Bhumali. They are also called 'Halalkhor'. There are certain other castes also doing the work of scavenging such as Doms, Dumras, Dhanuks, Bansphors etc. In Nepal sweeper caste include Pores, Chamis and Halahulas. But one thing is very clear that all of them stand at the very bottom of the social ladder and are regarded as very unclean both by the Hindus and Muslims. Col. Tod calls them the very refuse of mankind.

W. Crooke (1896) writing about this community states that the name Mehtar was commonly applied to the servants of the Emperor Humayun. Another title for them is Halalkhor, one who eats what is lawful, one whose earnings are legitimate. This euphemistic title is said to have been introduced by the Emperor Akbar. They are also known as Khakrob, or "sweeper of dust" and Baharwala, one who is not admitted into the house. Another euphemistic name for them in the Punjab is Musalli, "one who prays". From their religion and patron saint they are sometimes known, collectively, as Lalbegi, which

* Annexure II.I
** Annexure I.2
*** Annexure II.2

is really the name for one of their sub-castes.

Crooke's ethnographic notes on 'the Tribes and Castes of the North-Western India' are based to a large extent on the *'Knights of the Broom'* by R. Greeven. He states that "the modern Bhangi is apparently the representative of the Chandala of Manu, who is said to be descended by a Sudra from a Brahmani women. He ordains that they must live without the town, whence the name Antavasin or Antevasin, 'one who dwells near the boundaries'. Their sole wealth must be dogs and asses; their clothes must consist of the cerecloths of the dead; their dishes must be broken pots, and their ornaments of rustly iron. No one who regards his duties must hold any intercourse with them, and they must marry only among themselves—a prohibition which takes us back to the very beginning of the caste system. By day they may roam about for the purpose of work, be distinguished by the badges of the Raja, and they must carry out the corpse of any one who dies without kindred. They should always be employed to slay those who by the law are sentenced to be put to death, and they may take the clothes of the slain, their beds and their ornaments. The term Chandal is now-a-days used only in the sense of contumely, and the so called Chandals of Bengal invariably call themselves Namasudra, and "with characteristic jealousy the higher divisions of the caste apply the name Chandal to the lower, who in their turn pass it on to the Dom". The word Chandala, which, if it really comes from an Aryan root, may be connected with, Chanda, in the sense of 'evil or mischievous', was possibly the designation of some of the meaner non-Aryan or Dravidian race who were at an early time reduced to servitude, and compelled to perform the vilest functions of the Aryan commonwealth, but the term Bhangi can be applied to any definite ethnological unit is more than doubtful. Many of the special duties of the Chandala of Manu, such as the conveyance of corpses and the task of acting as public executioners, are now vested in the Dom and his kindred, with whom the Bhangi, as we now see him, is doubtless closely allied. But the modern names seem to imply that the present organisation of the caste may have been contemporaneous with the early Muhammadan conquest, and there seems reason to believe that the tribe, as we now find it, is made up of a number of different elements. This is corroborated by the divergent physical appearance of the race. Some Bhangis have the dark complexion, stunted figure, and peculiar dark flashing eye which is so characteristic of the Dom. Others, again, are of a much taller form and fair complexion, This may be perhaps accounted for partly by the fact that their admittance as servants into the higher class families facilitates illicit connection with superior race, and partly that the tribe habitually recruits itself by the admission of outcastes from the superior tribes. It has also been suggested that the names of some of their sub-castes point to the supposition that the caste may be made up of menials attached to various Rajput, Jat, or Musalman tribes; the Haris, with the Haras; the Dhe with the Dhe Jats; and the Rawats with the higher tribe of the same name. But of this there is no distinct evidence".

About their caste organisation, Crooke states that "the last census classifies them under five main sub-castes: Balmiki, derived from the tribal saint whose legends have been already given; Dhanuk, which though allied to the Bhangis, has been treated as a distinct tribe; Hela, Lal Begi, and Patharphor, or 'Stone-breaker'. One list from Benares divides the caste into nine endogamous sub-castes, —Shaikh, Hela, Lal Begi, Ghazipur Rawat, who trace their origin from Ghazipur, and take their name from the Sanskrit raja-duta,or "royal messengers" Hanri or Hari, who appear to be so called because they pick up bones (Sanskrit *hadda*) and other rubbish, Dhanuk, Bansphor and Dhe. Of these according to the Benares account the Lal Begis have their headquarters at Amritsar and Delhi; the Rawats at Agra, Mainpuri, Meerut, Ghazipur and Dinapur; the Shaikhs at Mirzapur and Delhi, and the Helas at Calcutta. The detailed census lists supply no less than thirteen hundred and fifty-nine sub-castes of Hindu and forty-seven of Mohammedan Bhangis."

Of the Benares sweepers, Greeven writes: "in Benares, only the Lal Begi, Shaikh Mehtar and Hela, with a few Rawats, are found. All sub-castes, including Lal Begis, who acknowledge a Musalman hero, claim to be Hindus, with the exception to the Shaikh Mehtars who call themselves Mohammedans. These pretensions are, however, equally rejected by Hindus, who exclude them from temples and by Musalmans who exclude them from mosques. The distinction between Lal Begis and Shaikh Mehtars is purely religious, and an elaborate legend admitting the common origin has been invented to explain why Mazhabis, who are Lal Begis converted to Nanakshahi doctrines, do not object to eating with Shaikh Mehtars. Only Lal Begis and Rawats eat food left by Europeans, but all eat food left either by Hindus or Musalmans".

Further he goes on to say: "the religion of the sweepers is a curious mixture of various faiths. Some, as we have seen, profess to be Hindus, others Musalmans, and other Sikhs, But though these two latter religions avowedly preach the equality of all men, they refuse to recognise sweepers as brethren in the faith. In Benares the Rawats are said to be as bad Hindus as the Shaikhs are indifferent Mohammedans, and the Chaudhari of Helas could say only that he professed the Hela religion. But the experience of the last Punjab census has shown the impossibility of classing their beliefs under any one definite creed. Some ninety five per cent of the Chuhras of the Province did, it is true record themselves as professing some religion which might be assumed to be peculiar to them, such as Lal Begi, Balmiki or Balashahi; but, as Mr. Maclagan observes: "While there is no doubt that we should be complying with Hindu feeling in excluding the Chuhra from the list of Hindus, should we also exclude the Chamar? And, if the Chamar, why not the Sansi? And should the Gagra, the Megh, and the Khatik follow? And, in fact where is the line to be drawn? In the absence of any clear decision on this point, it will be best to adhere to the present system and include all as Hindus". At the last census of these Provinces, 2,65,967 persons recorded themselves as votaries of Lal Begi. To the east of the province many are worshippers of the Panchon Pir. To the West Shikh Saddu and Gurunanak are worshipped. Guga or Zahir Pir is again held in high respect by the sweepers of the Western Districts. They consider that he cures the blind, lunatics, and leapers, and has the power of bestowing offspring on barren wives. His shrine is a small, round building with a courtyard and flags hung from a neighbouring tree. On the shrine is laid a leaf platter containing a chip of the wood of the pilu tree *(Careya arborea)*, a flower of the Karil or caper bush, and some bajra millet. The tomb is then rubbed with sandalwood, and this substance is considered a cure for various diseases. A goat is sometimes offered at a neighbouring shrine known as Gorakhnath Ka Quila; and every Lal Begi erects in his house a standard *(nishan)* in the form of a trident *(trisul)* in honour of Zahir Pir. In the eastern parts of these provinces, where distance overcomes the zeal for pilgrimage, it is usual for the Bhangis to carry round the sacred symbol of the Pir in the month of *Bhadon*, and raise contributions".

"Ghazi Miyan, again, is a favourite object of worship by Bhangis. They have corrupted the standard legend of the saint into a mass of extraordinary hagiology. According to one version Mamal and her father Sarsa fled from Delhi to Ghazni on account of the tyranny of Prithvi Raja. There Salar Sahu married Mamal and Sarsa managed to persuade Sultan Mahmud to attack Prithvi Raja. His tomb at Bahraich is a favourite place of Bhangi pilgrimage. The Dafali priests of the tomb perform all the rites. One of them wears the figure of a horse on his waist; Others follow him in a wild dance, singing the praises of Shah Madar. All this is in commemoration of the marriage of Ghazi Miyan, which is said to have taken place the day before his martyrdom."

"Bhangis, again, have an army of local deities, such as in Lucknow, Kale Gora, Baram Gusain, Narsinha, and Buddhi Prasadi. They believe largely in various evil spirits, the Bhut, the Deo, the Bir, the Rakshasa, and the Churel. They observe, if Hindus, the festivals of the faith, such as the Dwali, Ghazi Miyan-ka-byah, the Basant, Id, and Muharram, which are all observed by the Lal Begis of Benares; while the Gazipuri Rawats celebrate the Pachainyan, the Diwali, the Dithwan, the Khichari, the Holi, and Ghazi Miyan-ka-byah. The Helas observe the Holi, the Moharram, and the marriage of Ghazi Miyan, and the Shaikh Mehtars, the last, with the ordinary feasts of Islam. The common oaths in use are *Parmeswar qasm,* 'and *Khuda qasm.* The Lal Begis also swear by their patron saint. They plaster a place with cow dung, place a vessel of water inside it with a copy of the genealogy *(Kursi),* and the person swearing faces the Ka'ba and swears with the book in his hand".

Writing about the social rules, Crooke says that "Among the Hindu Bhangis of Lucknow, the women cannot wear the bodice *(angiya),* Chemisette *(Kurti),* or gold ornaments, and do not bore the nose for a ring. Muhammadan Bhangi women do not wear gold ornaments or sky-blue *(asmani)* or lac bangles *(churi).* The use of brass ornaments is considered unlucky, but those of alloy are allowed. They prefer earthen to metal cooking vessels, and no Bhangi will plant the *ber* tree or the bamboo before his door. The elder brother cannot touch the wife of his younger brother, and he can eat with no women but his own sister. If he touches a Dom he must purify himself before doing any other work. He will not eat food touched by a Dom or Dhobi, and the husband and wife will not mention each other by their names. Of all tribes the Dom, though he is admitted to be akin to the Bhangi, is held in particular abhorrence. Their rules of food vary with the religion they profess. Thus, Shaikh Mehtars will not eat pork, and some of the Hindu Bhangis will not eat beef. The Helas profess to eat the bearings of only high caste Hindus. No Bhangi, it appears, will eat monkeys, uncloven footed animals, scaleless fish, crocodiles, lizards, snakes, jackals, rats or other vermin. The Lal Begis salute in the form Ram! Ram!, Yadallah!, and Hardam Allah! To elders, they say Salam or Satnamko! Brahmins they salute with Maharaj! Palagan. The Ghazipuri Rawats and Helas salute everybody with Ram! Ram! With the exception of Mussalmans, to whom they say Salam or bandagi; and palagan to Brahmins. Shaikh use the word Salam only".

Writing about their occupation, Crooke states that "their

occupation is to remove filth, to sweep the houses and roads, to play on the flute or tambourine (Shahnaidaf) at marriages and other social occasions. They also conduct what is called *roshanchauki* at marriages, or when solemn vows *(mannat)* are made. Some of them are noted for their musical ability. The Hela makes winnowing fans and sieves *(sup, chhalni),* and some of the Shaikhs are collectors and appliers of leeches. The Bansphor makes baskets, mats, etc. The Dhanuks are fowlers and watchmen. They serve in the bands of native princes, and their women are midwives, To the west of the Province the Dhe's, a class of Lal Begis, act as hangmen and killers of pariah dogs. The Dhanuks and Bansphors will not remove night-soil, and the Shaikhs will not do this work at public latrines. Their implements are the broom *(Jharu)* and the rib bone of an ox *(panja),* with which they scrape up filth. Many of them are the hereditary priests of Sitala, and arrange the offerings of pigs released at their shrine; others serve Bhumiya and similar local godlings. In some places Bhangis are true village menials and receive a patch of rent-free land or some allowances at harvest in return for their services. In our cities, particularly in places like Mirzapur, where they are not numerous, they are much given to combination among themselves. They resent the settlement of new members of the tribe and allot the houses of the residents into certain beats *(halqa, Ilaqa)* each of which is served by a Bhangi and his wife. They call the occupants of such houses their "parishioners" *(jajman)* and fiercely resent the intrusion of any strange Bhangi within the beat; in fact most of the cases which come before the council relate to dispute of this kind".

Speaking of the scavenger community of Punjab Ibbetson (1916) says: "Socially they are the lowest of the low, even lower perhaps than the vagrant Sansi and the gipsy Nat, and as a rule can hardly be said to stand even at the foot of the social ladder, though some sections of the clan have mounted the first one or two steps. Their hereditary occupation is scavenging, sweeping the houses and streets, working up, carrying to the fields, and distributing manure, and in cities and in village houses where the women are strictly secluded, removing night-soil. They alone of all classes keep those impure animals, pigs and fowls; and they and the leather-workers alone eat the flesh of animals that have died of disease or by natural death. Together with the vagrants and gipsies they are the hereditary workers in grass and reeds, from which they make winnowing pans and other articles used in agriculture".

He further adds that "the Chuhra or Bhangi is the sweeper and scavenger *par excellence* of the Punjab is found throughout the

Province except in the hills. In the forntier Chuhras and Kutanas who have returned themselves as Jats. He is one of the village menials proper, who receive a customary share of the produce and perform certain duties. In the east of the Province he sweeps the houses and village, collects the cowdung, pats it into cakes and stacks it, works up the manure, helps with the cattle, and takes them from village to village. News of a death sent to friends is invariably carried by him and he is the general village messenger. He also makes the *chhaj* or winnowing pan, and the *sirki* or grass thatch used to cover carts and the like. In the centre of the Province he adds to these functions actual hard work at the plough and in the field".

"The Chuhras who have become Sikhs are called Mazbi and Rangreta. They often marry with the Lal Begi or Hindu Chuhra. They make capital soldiers and some of our regiments are wholly composed of Mazbis. The Rangreta are a class of Mazbi apparently found only in Ambala, Ludhiana and the neighbourhood, who consider themselves socially superior to the rest. Almost all the Chuhras west of Lahore are Musalmans, and they are very commonly called Musalli or Kutana, the two terms being apparently almost synonymous, but Kutana being used in the south-west and Musalli in the north-west." H.A. Rose in the Glossary of the Tribes and Castes of the Punjab and North-West Frontier Provinces (1919) states that, "Mazbi means nothing more than a member of the scavenger class converted to Sikhism. The Mazbis take the pahul, wear their hair log, and abstain from tobacco, and they apparently refuse to touch night-soil, though performing all the other offices hereditary to the Chuhra caste. Their great guru is Tegh Bahadur, whose mutilated body was brought back from Delhi by Chuhras who were then and there admitted to the faith by Guru Gobind as a reward for their devotion. But though good Sikhs so far as religious observance is concerned, the taint of hereditary pollution is upon them, and Sikhs of other castes refuse to associate with them even in religious ceremonies".

"One of the bravest of the generals of the Gurus, was Jiwan Singh, a Mazbi, whose tomb is still shown at Chamkaur in Ambala. He fell at its siege in 1705-06. During the Muhammadan persecution of the Sikhs they dropped out of notice and failing a supporter in the place of Guru Gobind, they never came to the front as a class, although Majaraja Ranjit Singh had a great admiration for their bravery and enlisted them freely. Being afraid, however, to form them into separate corps, he attached a company to various battalions. They were, however, looked down upon by the other men and naturally became discontented. When the Punjab was annexed, the Mazbi was a dacoit, a robber and often a *thug*. In this capacity he was generally styled a Rangretha. The latter are a class of Mazbi apparently found only in Ambala, Ludhiana, and the neighbourhood who consider themselves socially superior to the rest. The origin of their superiority, according to Sir Denzil Ibbetson's information, lies in the fact that they were once notorious as highway robbers! But it appears that the Rangrethas have very generally abandoned scavenging for leather-work, and this would at once account for their rise in the social scale. In the hills Rangretha is often used as synonymous with Rangrez, or Chhimba or Lilari, to denote the cotton dyer and stamper, and in Sirsa the Sikhs will often call any Chuhra whom they wish to please Rangretha, and a rhyme is current Rangretha, Guru Ka Beta, or "the Rangretha, is the son of the Guru". The Mazbis have social distinctions among themselves. The descendants of the true Mazbis who rescued Tegh Bahadur's body are strictly speaking, the only *asl* or real Mazbis, but the term is applied loosely to more recent converts. Recent converts are looked upon more or less with a critical eye and are termed Malwais. This term was probably a geographical distinction at first, but is now merely a caste one. It takes some generations to make a Mazbi. but how many he cannot say. Much depends on circumstances, and on the strictness of the convert's adherence to the faith as to when he may be admitted to an equal footing with a true Mazbi. For this reason the *asl* Mazbi is scarce and his physique is falling off. Until quite lately he was never found in large numbers in any special locality except for the purpose of work on a new canal or railway. Two or three Mazbi houses are attached to Jat villages where they work as labourers. Grants of land have, however, been made in Gujranwala to pensioners of Pioneer regiments. The Mazbi *gotras* are numerous and many of them are the same as those of the Jat, doubtless following the family or group whose hereditary servants they were. In their customs too, at weddings, etc., they conform to a great extent to those prevalent among the Jats".

Scavenging Castes of Central India

Writing about the Tribes and Castes of the Central Provinces of India (1916) Russel and Hiralal states that "in this province they generally confine themselves to their hereditary occupation of scavenging, and are rarely met with outside the towns and large villages. In most localities the supply of sweepers does not meet the demand. The case is quite different in northern India, where the sweepers castes—the Chuhra in the Punjab, the Bhangi in United Provinces and the Dom in Bengal—are all of them of great numerical strength. With these castes only a small proportion are employed on scavengers" work and the rest are labourers like the Chamars and Mahars of the Central Provinces. The present sweeper caste is made up of diverse elements and the name Mehtar, generally applied to it, is a title meaning a prince or leader. Its application to the caste, the most abject and despised in the Hindu community, is perhaps, partly ironical; but all the low castes have honorific titles, which are used as a method of address either from principle, as the Hindus say, that you may call an ass your uncle if you want him to do something for you. The regular caste of sweepers in northern India are the Bhangis, whose name is derived by Mr. Crooke from the Sanskrit *bhanga,* hemp, in allusion to the drunken habits of the caste. In support of this derivation he advances the Beria custom of calling their leaders Bhangi or hemp—drinker as a title of honour. In Mr. Greeven's account also, Lal Beg, the patron saint of the sweepers, is described as intoxicated with the hemp drug on two occasions. Mr. Bhimbhai Kirparam suggests that Bhangia means broken, and is applied to the sweepers because they split bamboos. In Kaira, he states, the regular trade of the Bhangias is the plaiting of baskets and other articles of split bamboos, and in that part of Gujarat if a Koli is asked to split a bamboo he will say, "Am I to do Bhangia's work? The derivation from the hemp-plant is, however, the more probable. In the Punjab, sweepers are known as Chuhara, and this name has been derived from their business of collecting and sweeping up scraps *(chura-jhurna).* Similarly, in Bombay they are known as Olganas or scrap-eaters. The Bengal name Hari is supposed to come from *haddi,* a bone; the Hari is the bone-gatherer, and was familiar to early settlers of Calcutta under the quaint designation of the 'harry-wench'. In the Central Provinces sections of the Ghasia, Mahar and Dom castes will do sweepers's work and are therefore amalgamated with the Mehtars. The caste is thus mixed constitution, and also forms a refuge for persons, expelled from their own societies for social offences. But though called by different names, the sweeper community in most provinces appear to have the same stock of traditions and legends. The name of Mehtar is now generally employed, and has therefore been taken as the designation of the caste."

Russel and Hiralal give a brief account of the occupational pattern of sweepers. They have recorded that "sweeping and scavenging in the streets and in private houses are the traditional occupations of the caste, but they have others. In Bombay they serve as night watchmen. town-criers, drummers, trumpeters and hangmen. Formerly the office of hangman was confined to sweepers, but now many low-caste prisioners are willing to undertake it for the sake of the privilege of smoking tobacco in jail which it confers. In the Central Provinces the hangman was accompanied by four or five other sweepers of the caste panchayat, the idea being perhaps that his act should be condoned by their presence and approval and he should escape guilt. The hangman received ten rupees as his fee, and of this five rupees were given to the caste for a feast and an offering to Lal Beg expiate his sin. Sweepers are, as a rule, to be found only in large villages, as in small ones there is no work for them. The caste is none too numerous in the Central Provinces, and in villages the sweeper is often not available when wanted for cleaning the streets. The Chamars of Bundelkhand will not remove the corpses of a cat or a dog or a squirrel, and a sweeper must be obtained for the purpose. These three animals are in a manner holy, and it is considered a sin to kill any one of them. But their corpses are unclean. A Chamar also refuses to touch the corpse of a donkey, but a Kumhar (potter) will sometimes do this; if he declines a sweeper must be fetched. When a sweeper has to enter a house in order to take out the body of an animal, it is cleaned and whitewashed after he has been in. In Hoshangabad an objection appears to be felt to the entry of a sweeper by the door, as it is stated that a ladder is placed for him, so that he presumably climbs through a window. Or where there are no windows it is possible that the ladder may protect the sacred thereshold from contact with his feet. The sweeper also attends at funerals and assist to prepare the pyre; he receives the winding-sheet when this is not burnt or buried with the corpse, and the copper coins which are left on the ground as purchase-money for the site of the grave. In Bombay in rich families the winding-sheet is often a worked shawl costing from fifty to a hundred rupees. When a Hindu widow breaks her bangles after her husband's death, she gives them, including one or two whole ones, to a Bhangia woman. A letter announcing a death is always carried by a sweeper. In Bengal a funeral could not be held without the presence of a Dom, whose functions are described by Mr. Sheering as follows: "on the arrival of the dead body at the place of cremation, which in Benares is at the basis of one of the steep stairs or *ghats,* called the Burning Ghat, leading down from the streets above to the bed of the river Ganges, the Dom supplies five logs of wood, which he lays in order upon the ground, the rest of the wood being given by the family of the deceased. When the pile is ready for burning a handful of lighted straw is brought by the Dom, and is taken from him and applied by one of the chief members of the family to the wood. The Dom is the only person who can furnish the lights for the purpose; and if for any reason no Dom is available, great delay and inconvenience are apt to arise. The Dom exacts his fee for three things, namely, first for the five logs, secondly for the bunch of straw, and thirdly for the light".

It is a further recorded that "during an eclipse the sweepers reap a good harvest; for it is believed that Rahu, the demon who devours the sun and moon and thus causes an eclipse, was either a sweeper or the deity of the sweepers, and alms given to them at this time will appease him and cause him to let the luminaries go. Or, according to another account, the sun and moon are in Rahu's debt, and he comes and duns them, and this is the eclipse; and the alms given to sweepers are a means of paying the debt. In Gujarat as soon as the darkening sets in, the Bhangis go about shouting, *Garhandan, Vastradan, Rupadan*' or gifts for the eclipse, gifts of clothes, gifts of silver.

Scavenging Castes of Western India

The Bhangis are divided into different social units at various levels such as territory, religion, occupation, descent, etc. R.E Enthoven in his *Tribes and Castes of Bombay* (1920) reports that "Bhangis or Bhangias are found in almost every district of the Presidency and Native states. In Gujarat, they are known as Halalkhors, Olgana Barvashia, Metariya, Jamphoda and Mela and in the Deccan and Karnataka as Halalkhors. As a caste of scavengers and sweepers they are the dregs of Hindu society, and contain an admixture of outcastes who have fallen to this level owing to offences against the social code of higher castes. Being open to continual recruitment in this fashion their customs are confused and uncertain. Bhangia is generally supposed to mean broken".

"In the Deccan, the Bhangis are divided into Bhasods, Chajgadis, Helas, Lal Begis, Makhiyars and Sheikhs. Of these, Lal Begis and Sheikhs eat together but do not inter-marry and are considered higher than the other four, who do not inter-marry or eat together. Sheikhs profess to be Musalmans, and Lal Begis are half Hindus and half Musalmans. Many Bhangis in the northern part of the Presidency appear to be immigrants from the United Provinces. It seems probable that in many cases Bhangias originally came to this Presidency as camp followers with the armies from the north".

According to Enthoven "Most Bhangias, both men and women, are scavengers and night-carriers. They also sweep the roads, winnowing the dust in the hope of finding fragments of gold and silver, make baskets and do other bamboo work, and bury dead animals, cattle excepted. Some serve as trackers, messengers and letter carriers.

A letter telling of a death was formerly brought by Bhangia. They also serve as night-watchmen, town-criers, drummers, trumpeters, and hangmen. A few Bhangias cultivate in addition to their regular work. In North Gujarat, except the dragging away of dead cattle, all menial village work falls on the Bhangia. Besides sweeping the roads and carrying away all dead animals, except cattle, the Bhangias watches, shews the road, arranges, for supplies and points out boundaries. In municipal towns as scavengers men earn Rs. 8 to Rs. 20 a month and women Rs. 5 to Rs. 10."

In Goa the 'Bhangui' are represented in a very small number. The ancestors of the Bhangui of Goa have migrated mostly from the Bombay Presidency and Rajasthan about a hundred years ago. They are found mainly in Panaji. During 1971 census, there were the Bhanguis in Goa. They constitute an endogamous group. The division of the community is through the surnames such as Waghele, Parmar, Multani, Goher, Chavan, Nandrekar, Chatribin and Chavaria etc. Many of them have given up their traditional occupation of scavenging.

In Rajasthan, the Bhangis use clan name as their titles. Their women are called Bhangan by the local people. Munshi Hardyal Singh in his *The Castes of Marwar* (1894) writes that the Bhangis or sweepers are the lowest caste and are regarded as very unclean by the Hindus. They are employed in sweeping and removing night-soil. They also make *chhaj* or winnowing pans. They are very rarely found in the country, and abound chiefly in large towns. The Bhangis are also called Lal Begis, Khakrobs, Halalkhors and Mehtars, which name are thus accounted for by Sir Elliot: "Bhangi probably from their drunken habits, from bhang; Lal Begi from their object of worship; Khakrob from persian *Khak,* earth and rob sweeping Halalkhor, from Persian *Halal,* lawful, and *khor,* eating, because every thing is lawful to them as food, Mehtar from Persian Mehtar prince, said to have been applied to them in derison".

He further adds that "in Marwar there are two divisions of the Bhangis viz the Loharis and the Multanis who intermarry with each other. The chief peculiarity with the Bhangis of Marwar is that they will never eat food, if invited in a public feast, though they will readily accept the leavings of other guests, It is said that when on the demise of the late Maharaja Takht Singh, all the people of Marwar were fed (the ceremony being known as Sher-Sarni) the Bhangis alone could not be induced to take part in it unless offered with the golden broom and basket or *jharu-tokra*".

Scavenging Castes of Eastern India

Hari, Mehtar, Har-Santan and scavenger caste of West Bengal proper, which Dr. Wise identifies with the Bhuinmali and regards as "the remnant of a Hinduised aboriginal tribe which was driven into Bengal by the Aryans or the persecuting Mohammedans". H.H. Risley (1891) records that "This conjecture is borne out by the physical features of the caste, and by the fact that wherever we can trace the affinities the menial castes we find them to have been largely recruited, if not entirely drawn, from the ranks of the aboriginal races. The internal structure of the Hari caste throws no light upon its origin, as at the present day there are no sections, and marriage is regulated solely by counting prohibited degrees. The sub-castes are the following: Bara-bhagiya or Kaora-paik, Madhya-bhagiya or Madhaukul, Khore or Khoriya, Siuli, Mehtar, Bangali, Maghaya, Karaiya, Parandwar. Of these, the Mehtar sub-caste alone are employed in removing night-soil; the Barabhagiya serve as chaukidars, musicians, and palki-bearers; the Khore keep pigs; the Suili tap date-palms for their juice, and the rest cultivate. Some authorities include Ghasi among the sub-castes of Hari, but I have preferred to treat them as a separate caste".

Risley further records that "Their social rank is of the lowest. No one will eat with a Hari or take water from his hands, and members of the caste are not permitted to enter the courtyard of the great temples. Some of them hold land as occupancy or non-occupancy raiyats, and many find employment as agricultural day-labourers. Tapping date-trees, making bamboo combs, playing musical instruments at weddings and festivals, carrying palanquins, serving as syees, and scavenging, are among their characteristic occupations; but the removal of night-soil is confined to the Mehtar sub-caste. their women often act as midwives. They are troubled by few scruples regarding diet. Fowls, pork, field-rats, scaly or scaleless fish, and the leavings of other castes, are freely eaten; while in district where the aboriginal races are numerous Hari will even indulge in beef. Their partiality for strong drink is notorious".

In Assam, the sweepers and scavengers are known as Bansphor, Lalbegi, Hari, Maghaia, Dom, Ghashi, Uria, Goal Tanti, Hela, Kurmi Turi, Koiree, Goar, Kakar, Bagti, Sammar, Kol, Sheck, Chuna, Naga, Kubui Naga, Shekra, Raut, Mehtor, Sumar and Kurmai. They consider the saint, Balmiki, as their eponymous ancestor and thus introduce themselves as Balmiki. They are mostly distributed in and around urban centres. They are divided into clans, such as Pawar, Dhekia Pawar, Tejia, Sarowan, Chasria etc. Their traditional occupation is sweeping and scavenging. A few families in Jorhat (Kalyanpur) have started cultivation on a small scale. Some of them have adopted the jobs of plaiting baskets and making other articles of split bamboo like brooms etc. According to the 'Cosmogony of Caste and Social Mobility in Assam by Bimal J. Dev & Dilit K. Lahiri (1984)' "Bansphor" was the most dominant of all the caste groups among the sweepers and was found in most of the districts of Assam. A significant aspect is that a number of ex-tea-garden labour castes got involved in this profession especially in the Jorhat sub-division mainly out of economic distress. Consequently, they had to suffer social ostracism and rest content with a degraded status. In the district of Cachar there still prevails a class of sweepers belonging to 'Chuna Naga' and 'Kabui Naga' groups who at one time fled to Cachar from Manipur in view of the oppressive dealings of the Manipur State Darbar. Moreover, some of the earlier sweeper groups seem to have risen higher in the social scale through occupational diversification. This is illustrated by the vertical mobility of Bhuimalis in the Surma Valley and Haris in the Assam valley."

Scavenging Castes of Southern India

In Southern parts of the country most of the sweepers and scavengers are migrants from the Northern India, with the exception of Paky, Paraiyans, Sanei, Ghasi, Mala Korar, Malkana or Zadamalli etc. There are others also engaged in this profession like Lingayats, Kuruba, Mudaliars, Adi Andhras, Adi-Dravidas and Adi-Karnatakas etc. Syed Serajul Hassan* (1920) writes that "the Lal Begis of the Nizam's dominions have seven groups, who neither eat together nor intermarry. These are the Hele, Malkane, Chhichhade, Dumar, Chhajgode, Bhadeye Dassan and Makhyar. They also have a number of exogamous groups, such as Phatrod, Sanakat, Kanderas, Surawals, Gaikwad, Sarwan and Kanderia. Their traditional occupation of carrying night-soil is now being abandoned. They are engaged as sweepers and scavengers and most of them are employed in the municipalities".

---

* *The Castes and Tribes of H.E.H. the Nizam's Dominions* by Syed Siraj-Ul-Hassan.

3

INITIATIVES TAKEN BY THE CENTRAL AND STATE GOVERNMENTS SINCE INDEPENDENCE

Before the advent of Mahatma Gandhi nobody took up the cause of liberation of scavengers from their hereditary, inhuman and disgraceful occupation of manually removing human excreta and filth to earn their livelihood. Mahatma Gandhi had realised the situation in the early twenties and said:

> "I do not want to attain *Moksha,* I do not want to be reborn. But if I have to be reborn, I should be born an untouchable, so that I may share their sorrows, sufferings and the affronts levelled at them, in order that I may endeavour to free myself and them from that miserable condition. I therefore, pray that I should be born again, I should do so not as a Brahmin, Kshatriya, Vaishya or Shudra but as an Atishudra a Bhangi".

In the Ashram under the tender care and guidance of Mahatma Gandhi, the scavenging work was performed by the Ashramites of all castes themselves and dignity was added to the profession.

Barve Committee

The problem of scavenging and improving the conditions of scavengers has persistently been engaging the attention of the Government since independence. The erstwhile Government of Bombay, appointed a committee known as the Scavengers' Living Conditions Enquiry Committee in 1949 with late Shri V.N. Barve as Chairman to study and enquire into the living conditions of the scavengers in the State of Bombay and to suggest ways and means to improve their present conditions of work and to fix their minimum wages. The Committee submitted its report to the Government of Bombay in 1952.

In 1955 the Ministry of Home Affairs circulated a copy of the major recommendations* of the Barve Committee to all the State Governments requesting them to adopt these recommendations as they were really capable of wider application and could be implemented profitably by all State Governments. A similar action was taken by the Ministry of Health also.

Scheme for the Supply of Wheel-barrows and Improved Implements to Scavengers

The Government of India, as a first step to make the work of scavenging consistent with human dignity, suggested supplying of wheel-barrows/hand carts to sweepers employed by municipalities etc for scavenging work so that the practice of carrying night-soil as head load could be eliminated altogether. As it was felt that this requirement if left alone to the municipalities etc might not, however, be fulfilled in the absence of adequate financial resources, the Government of India in the Ministry of Home Affairs in 1957-58 offered to subscribe 50% of the cost towards the purchase of wheel-barows/hand carts through the State Governments to such of the needy municipalities and other local bodies as might undertake to provide immediately to scavengers employed on this type of work, wheel-barrows or hand carts to eliminate the practice of carrying nightsoil as head load and were prepared to contribute the remaining 50% cost with or without the assistance from the State Governments and the grant-in-aid would be admissible only to such local bodies as agreed to stop this practice completely and not those taking up the scheme piecemeal. But the response from the local bodies/States was rather very poor. In many cases the amount sanctioned to the State Governments were not utilised at all, and in cases where amounts have been utilised these were negligible. The wheel-barrows provided to the scavengers were most improper and so heavy that they were bound to be discarded by the scavengers.

Backward Classes Commission (Kaka Kalelkar Commission)

The first Backward Classes Commission which was appointed in 1953

* Annexure III.I

under the Chairmanship of Kaka Kalelkar submitted its report in 1955. The Commission described the condition of sweepers and scavengers as sub-human. The Commission observed that "the alert community of Harijans has compelled us to look into their condition although technically the problem is not within our terms of reference. We visited Harijan quarters and studied their situation. The term "sub-human" can alone appropriately describe their condition. Bhangis cleanse our latrines and help maintain some measure of health and sanitation. Without Bhangis, the whole population would have to face the ravage of epidemics. And yet, these very Bhangis are forced to live in the filthiest of surroundings, and in some places they are even asked to carry nightsoil on their heads. The receptacles for nightsoil should not leak. Any society that expects one section of humanity to perform such humiliating and degrading service should be ashamed of itself. Municipalities are the greatest sinners in this respect. The plea that they do not have enough funds to improve the quarters of the Bhangis is hollow. If the city fathers had the same feelings for these unfortunates as they seem to have for their clerks and chaprasis, they would have managed to find resources to house the Bhangis in decent quarters. Mahatmaji had to threaten a fast in order to secure permission for one of his Brahman colleagues of the Ashram to clean the latrines of a British jail. Let all city fathers who love to call Mahatma Gandhi 'Father of the Nation', do something concrete in his manner to remove this blot from our civilization. The Bhangis should not be condemned to live in segregated localities. They should be distributed and given quarters among other groups".

The observations and recommendations of the Backward Classes Commission were brought to the notice of all the State Governments by the Ministry of Home Affairs in October 1956 emphasising the need to introduce mechanical and up-to-date methods of cleansing latrines so that the inhuman practice of doing this work by hand and carrying nightsoil on heads is obviated as far as possible and also specific schemes covering every sphere of life to uplift Bhangis from their 'sub-human' level of existence.

Central Advisory Board for Harijan Welfare

The Ministry of Home Affairs constituted a Central Advisory Board of Harijan Welfare in 1956 under the chairmanship of Late Pandit Gobind Ballabh Pant, Minister of Home Affairs. This Board *inter alia* reviewed the working and living condition of the sweepers and scavengers in the country and recommended to the Government to introduce a Centrally Sponsored Scheme for this purpose.

Malkani Committee

The Board in its meeting held on the 12th October, 1957 constituted a committee, known as Scavenging Conditions Inquiry Committee consisting of Prof. N.R. Malkani as Chairman and Sarvashri K.L. Balmiki, R.K. Bose, N.S. Kajrolkar and P.N. Rajabhoj as Members to prepare a scheme to put an end to the degrading practice of scavenging having to carry nightsoil in buckets or baskets. The Committee started work in January 1958 and submitted its report in December 1960. The report is informed by a genuine feeling for the scavenger's lot and contains concrete recommendations* not merely for eliminating the practice of carrying nightsoil as head loads, but also for removing filth and indignity from all stages of scavenging and for improving the working and living conditions and social status. Some of the recommendations made by the Committee are as follows :

(A) Working Conditions

(i) For putting an end to the practice of carrying night-soil as head load or waste load, wheel-barrows should be introduced.

(ii) The use of receptacles in private and public latrines should be made compulsory.

(iii) It should be obligatory to use a standardised scraper which should be supplied by the local bodies to their employees and not left to the scavengers themselves.

(iv) Efforts should be made to encourage the use of rubber gloves, particularly at the pail-depots and dumping grounds.

(v) Carrying of night-soil in basket, tasla and drum etc., as head load should be prohibited by law and made an offence.

(B) Housing for Sweepers and Scavengers

(i) Construction of quarters of conservancy staff should be near their place of work as far as possible but never around the pail-depots, morgues, public latrines, dumping grounds etc. These quarters should be built in colonies inhabited by others so as to avoid segregation.

(ii) The housing of scavengers should be improved so as to create

---

* Annexure III.2.

a new sense of self-respect amongst scavengers particularly their children.

(iii) One-room tenements would not be adequate as scavengers have large families. From the point of view of moral hygiene every tenements must have at least two rooms, a kitchen, a bathroom, a latrine and also if possible electricity and water supply.

(C) Wages

(i) Wages should be paid to scavengers according to the Minimum Wages Act. State Governments who have not done it so far should appoint committees to go into the question of wages and payment of other allowances to scavengers and suggest basic wages which should, however, not fall below the minimum wages.

(ii) Unclean work allowance should not be in cash but in the form of extra amenities such as gloves, availability of water etc.

(iii) For full-time employees the hours of work should not be more than 7 hours per day of which only four to five hours be reserved for scavenging work. Part-time workers should not be asked to work both in the morning and evening as this deprives them of the chance of obtaining other work.

(iv) Local bodies should frame leave rules and scavengers should be given the benefit of casual, privileged and earned leave etc.

(v) State Government should see that at least in the local bodies where scavenging service is municipalised, scavengers are allowed a full day off in a week.

The Committee observed that "as long as dry latrines continue, the problem will exist and it is a story of the far distant future to think of all the towns having underground sewers".

The various recommendations/observations of the Committee were sent to all the State Governments by the Ministry of Home Affairs for implementation. A Centrally Sponsored Scheme for improving the working and living conditions of sweepers and scavengers was also introduced during the Third Five Year Plan which will be discussed in the preceding paragraphs.

Committee on Customary Rights

The Central Department of Social Welfare appointed a committee in

1965 under the chairmanship of Prof. N.R. Malkani, to examine the question of the abolition of customary rights of the scavengers. The Committee submitted* its report to the Government in 1966. The Committee found that where scavenging is not municipalised the latrines were cleaned privately and one particular scavenger acquired hereditary rights to clean such latrine as against another scavenger by an understanding and agreement. A customary relationship also develops with the householder and the scavenger receives payments in some form or the other.

The response of the Government of India as well as the State Governments to various recommendations of this Committee was very poor and hence no action could be taken.

Pandya Committee

The National Commission on Labour constituted by the Union Ministry of Labour appointed a sub-committee (1968-69) under the chairmanship of Shri Bhanu Prasad Pandya to look into the working and service conditions of sweepers and scavengers which also met the same fate at the hands of the Central and State Governments as happened in the case of the report on Customary Rights. One of the important recommendations of the committee was that "the Central Government should undertake a comprehensive legislation for regulating their working, service and living conditions which should also provide for adequate inspectorate and enforcement machinery".

Similar committees were also constituted at the State level in the States of Uttar Pradesh (1955), Haryana (1969), Kerala (1971) and Karnataka (1976).

Improvement in the Working and Living Conditions of those engaged in Unclean Occupations

As a follow up of the various recommendations made by the Malkani Committee, the Government of India in the Ministry of Home Affairs introduced a scheme during the Third Five Year Plan as a centrally sponsored scheme which consisted of two parts viz. (i) improvement in the working conditions of sweepers, scavengers, tanners and flayers, and (ii) improvement in the living conditions of those Scheduled Castes who were engaged in unclean occupations or were landless labourers. The financial assistance was provided by the Government

* Annexure III.3.

of India to the State Governments so that they could give (a) grant-in-aid to municipalities/local bodies for the purchase of hand-carts, wheel-barrows, scrapers, gum-boots and the protective devices and (b) subsidy for construction of houses for sweepers, scavengers, tanners and flayers, and house-sites to members of Scheduled Castes who were engaged in unclean occupations or were landless labourers. This programme was only a supplementary one as the long term solution to the problem of scavenging depends upon the conversion of dry latrines into sanitary latrines and measures for improved conservancy and sanitation services like septic tanks etc.

Shortcomings of the Scheme

The scheme could not make much headway due to the following reasons:

- (i) The design of the wheel-barrows was not suitable for women and children who were the main users since it was too heavy to be handled by them.
- (ii) No financial assistance was provided to the Municipalities for maintenance and repair of the wheel-barrows.
- (iii) There was no arrangement for parking of the wheel-barrows.
- (iv) The improved implements like receptacles, scrapers, improved brooms, brushes, gumboots and handgloves etc. were not supplied to all the scavengers, and, if supplied to some, it was seldom used by them.
- (v) It was noticed that in some municipalities 50 per cent contribution as envisaged in the scheme, was not provided by them.

The scheme proved to be a failure and was therefore discontinued during the Fifth Five Year Plan.

Gandhi Centenary Year

During the Gandhi Centenary Year (1969) the Government of India took up a special programme for converting dry latrines into water-pour flush latrines. Under this scheme the householders were provided 25 per cent subsidy and 75 per cent loan recoverable in easy instalments. The householders were also allowed to connect their service latrines to nearby public sewer, if any.

The Ministry of Health* had suggested to the local bodies and

* Annexure III.4.

municipalities:

(i) No new buildings should be allowed to be constructed without provision of flush-out latrines;
(ii) In areas where sewers have been laid, strict enforcement of law for compulsory connection of dry latrines in the existing houses to sewers should be launched; and
(iii) Where sewers are not laid, a special drive should be launched for conversion of dry latrines into flush-out latrines in selected areas by connecting them to local septic tanks or leaching pits.

Centrally Sponsored Scheme of Ministry of Works and Housing

The Ministry of Works and Housing took up a scheme* on pilot-cum-demonstration project for the conversion of dry latrines into water borne latrines in the selected towns during the Fifth Five Year Plan and provided an outlay of Rs. 4.40 crores for this purpose. Some 30 towns were selected in the country under this scheme. The financial assistance provided to the State Governments was in the form of loans. The scheme could not make much headway as there was no element of subsidy in the scheme; with the result that the response of the State Governments towards this scheme was very poor. The scheme was, therefore, dropped during the Sixth Five Year Plan.

Pre-Matric Scholarships for the Children of those engaged in Unclean Occupations viz. Sweepers, Scavengers, Tanners and Flayers

The Centrally Sponsored Scheme of Pre-Matric Scholarships for the children of those engaged in unclean occupations viz sweepers, scavengers, tanners and flayers is in operation since Sixth Five Year Plan. The main objective of the scheme is to provide good quality of education to the children of those engaged in unclean occupations. The scheme was originally intended for hostlers only but it has now been modified (1991-92) to cover day scholars also. Scholarship is provided to the children staying in the hostels at the rate of Rs. 200 per month per child for classes III to VIII and Rs. 250 per month per child for classes IX and X. The rate of scholarships for day scholars is Rs. 25 per month for classes I to V, Rs. 40 per month for VI to VIII

---

* Annexure III.5.

and Rs. 50 per month for IX and X. The children also get Rs. 500 per annum as an ad-hoc amount to meet other expenses.

The scheme has undergone substantial modifications to make more purposeful by removing income ceiling criteria, providing scholarship to all children up to class VIII with condition that if third child is born after April 1993, only two children of the family would be eligible for this scholarship and this restriction is relaxed upto 2 children in classes IX and X.

During 1995-96 central assistance of Rs. 892 lakhs was released to the State Governments under this scheme and an equal amount was provided by them. The allocation made for this scheme during 1996-97 was Rs.10 crores. The achievements under this scheme was as follows:

Year	*No. of students benefited*
1991-92	90,912
1992-93	99,912
1993-94	1,30,715
1994-95	1,76,253
1995-96	2,40,983 (Likely)

The major constraint in the way of educational development of sweepers and scavengers is the unwillingness on the part of parents to send their children to school. The unwillingness of the parents is due to economic reasons. The children help their parents in earning livelihood. The scheme is to induce the parents to send their children to schools and to compensate them for the loss of income on this account. The scheme has not made much headway so far as the coverage of children is still very small.

Role of Non-Government Organisations

Besides the emergence of Sulabh International Social Service Organisation as the most effective and dynamic voluntary organisation contributory to the liberation of scavengers, a few voluntary organisations have done pioneering work in this field. The most prominent amongst them are the Harijan Sevak Sangh; Safai Vidyalaya, Ahmedabad; and the Gandhi Smark Nidhi, Pune.

Harijan Sevak Sangh

The Harijan Sevak Sangh was founded on September 30, 1932 as a result of Mahatma Gandhi's epic fast in the Yeravada Jail near Poona. This fast was in protest against the Communal Award of the British Government which threatened to separate the so called untouchables from Hindu society. The object of the Sangh was to eradicate untouchability through truthful and non-violent means, to remove the difficulties and the disabilities of the untouchables and to secure for them a status of full equality in all fields with the other Hindus.

The Sangh had from the beginning sought the co-operation of the municipalities and other local bodies in ameliorating the working and living conditions of sweepers and scavengers. The response had been slow but a beginning had been made by drawing the attention of these bodies to the lot of the sweepers and scavengers. Thirty-five co-operative societies functioned under the Sangh during the third year of its existence (1935) to provide credit to them and save them from the clutches of money-lenders. In Bombay, 226 rooms had been hired by the Sangh and let out to the sweepers and scavengers at cheap rates. As a result of the Sangh's intervention the Ujjain and Bhilsa Municipalities in the Gwalior State began to give loans to sweepers at a low rate of interest. In UP, Faizabad Municipality advanced Rs. 500 to the Sweeper's Co-operative Society and agreed to recover the amount in monthly instalments from their salaries. In Meerut (UP) and Bellary (now in Karnataka), the Sangh was able to secure the benefit of maternity leave for women sweepers.

After the introduction of provincial autonomy in 1936, the Sangh made representations to Congress Ministries for bringing about improvements in the working and living conditions of sweepers employed by local bodies so that they might get a decent minimum wage, proper housing, education and other facilities and might be saved from the corrupt officials and money-lenders.

The Municipalities in Bombay, Ahmedabad, Madurai, Tiruchirapalli, Bangalore, Madras and Kanpur constructed residential quarters for their sweepers. The Sangh helped the co-operative society of sweepers at Ranavar in Kathiawar to build quarters for sweepers.

Six Sweeper's Sevaks were working for the welfare of the sweepers in different parts of the country under the new scheme started by the Sangh.

After independence the work of the Sangh was increased manifold. The Central Board of the Sangh in its meeting held on 15th and 16th November, 1958 discussed the hardships of the sweepers and scavengers and welcomed a suggestion by Shri Appa Saheb Patwardhan that a permanent exhibition be organised at the Headquarters of the Sangh in Delhi where models should be displayed of improved latrines and implements for cleaning and removal of night-soil. The Sangh had been receiving grants from the Central Government for Bhangi Kashta Mukti programme. The Sangh runs this programme through the Safai Vidyalaya at Ahmedabad. The 78 bhangi sevaks employed by the Sangh were working in 14 branches, contacting municipal authorities for the betterment of the working and living conditions of the sweepers and encouraging the sweepers themselves to adopt more scientific ways of work. They also helped in the formation of credit co-operative societies and 125 of these were functioning. There were in addition, a large number of industrial, housing and agricultural societies which they encouraged. As a result of the work of these bhangi sevaks, more than 7000 wheel-barrows were supplied to the scavengers and about 400 residential quarters constructed for them.

The Sangh felt that while emphasis is to be laid on improved methods of working for the sweepers, it was clear in its intention that the ultimate goal was to make it unnecessary for man to carry or clean human excreta and therefore the Sangh had been laying enphasis on better types of latrines. But, it was long cry to the day when even all the municipalities in the country, could claim that every single latrine within their jurisdiction had sanitary fittings. The attack was, therefore, two-pronged. On the one hand, the Sangh and its workers devoted great attention to the welfare of sweepers and improvement in their methods of work, at the same time it was trying to persuade the municipalities and other local bodies to make it obligatory for house-owners to have only sanitary latrines.

Safai Vidyalaya, Ahmedabad

Safai Vidyalaya, a sanitation institution was set up by the Harijan Sevak Sangh at Ahmedabad under the leadership of Shri Ishwar Bhai Patel and entrusted with the implementation of Bhangi Kashta Mukti Programme which provides training in respect of the method, approach, use of improved implements for cleaning latrines etc. to junior and senior engineers, masons, sweepers and scavengers besides sanitary inspectors of primary health centres in rural areas and social workers. Short-term courses for policy makers and administrators are also arranged at various levels to convince them of the importance of sanitation.

This Institution under the dynamic leadership of Shri Ishwar Bhai Patel has been very helpful in getting various bye-laws changed and subsidies granted by the local bodies and panchayats for conversion of dry latrines into pour-flush sanitary ones to eradicate the sub-human practice of carrying night-soil as head load by the scavengers. The available information shows that 1.31 lakh dry latrines were converted in Gujarat from 1964-65 to 1977-78 out of total of 1.75 lakhs dry latrines. By 1988, the total number of 1.80 lakh dry latrines were converted (1.05 lakh in urban areas and 0.75 lakh in rural areas) out of 1.86 lakh dry latrines which were increased from 1.75 lakhs due to discovery of new dry latrines. An evaluative study of low cost sanitation, in Gujarat (1991) revealed that most of the scavengers' children are unemployed inspite of their educational attainment at different levels. This has caused restlessness and tension in the scavenger community.

Gandhi Smarak Nidhi, Pune

The Gandhi Smarak Nidhi Pune has been working to abolish scavenging by converting the service latrines into pour-flush sanitary latrines and also by constructing new latrines which do not require the services of scavengers. The Nidhi had also set up a Cell—Bhangi Mukti Yojana—for propagation and construction of sanitary latrines. The Nidhi also imparts training for social and municipal workers which also includes training for the construction of sanitary latrines and biogas plant.

Sulabh International Social Service Organisation

One of India's most serious environmental hygiene problem is proper management and disposal of human waste. The neglect of this basic sanitary measure is a chronic feature in the development of most of our towns and villages and the cause of most water-borne diseases. At present according to the figures made available by the 1991 Census more than 76 per cent of the people in India have no toilet facility and only about 24 per cent have toilet facility. A large percentage of people in both rural and urban areas, therefore, defecate in any open space available. With the phenomenal rise in population, the problem has become more acute and it is necessary to provide toilet facilities to the common man at a cost which is affordable, socially acceptable and easily available.

The organisation that has come to the forefront in performing a sterling service in the field of low-cost public sanitation in both rural and urban areas is Sulabh International Social Service Organisation. Dr. Bindeshwar Pathak, Founder of the Sulabh Movement took up the challenge for the liberation of scavengers from carrying night-soil as head load during the Gandhi Centenary year in 1969. In 1970 he founded a non-profit voluntary social organisation named Sulabh Shauchalaya Sansthan (now known as Sulabh International Social Service Organisation) which was recognised by the Government of Bihar as the only organisation having the expertise of the leach-pit type latrines in the State and a catalytic agency to work between the beneficiaries, local bodies and the Government In 1968-69 the Government of Bihar made an allocation of Rs. 2 lakh to various local bodies including the Patna Municipal Corporation for the conversion of dry latrines into Sulabh Shauchalayas. It was because of the excellent work done by Dr. Pathak with missionary zeal in this field that the budget of the Government of Bihar was increased to Rs. 2 crores in 1985-86. The research and development of cheap and appropriate sanitation technology has been one of Dr. Pathak's most important programmes. The technology developed by him is cost-effective, consumes less water and requires material and know-how that is available easily and indigenously. This scavenging free-Sulabh technology got the backing of WHO, World Bank and UNDP etc. The United Nations Centre for Human Settlements recognised India's cost-effective and appropriate sanitation systems—Sulabh International system—as a global *'Urban Best Practices'* in the Habitat II conference held at Istanbul, Turkey in June, 1996. The Economic and Social Council of United Nations has recently granted *special consultative status* to this organisation.

Sulabh's Technology

Sustainability and replicability are the two basic principles of an appropriate technology which is fulfilled by the Sulabh Shauchalaya. Sulabh Shauchalaya (twin pit pour-flush toilet) developed* by Sulabh International in 1970 is socio-culturally acceptable, affordable, easily available and users are able to operate and maintain it easily. It provides all the health benefits by safe disposal of human excreta on site, which sewerage provides. Sulabh Shauchalaya is most appropriate to serve as an alternative to bucket privies and stop open air defecation. It can be constructed even in the most congested areas of cities and towns and has been widely accepted by the people because of its low-cost and requires only two litres of water to flush and can thus function even where enough water is not available. Neither it requires the services of scavenger nor it pollutes the air. The waste collected in the

* Annexure IV.1.

pit is converted into rich manure and good soil conditioner. The squattering pan can be cleaned and easily maintained by house owners themselves. The two pits work alternatively continuously. It has a high potential for upgradation i.e., can be easily connected to sewer when introduced in the area. Sulabh Shauchalayas can be built on upper floors of the houses too. It can be provided within the house as it has no foul smell or fly mosquito nuisance. In the case of such toilets, vent pipe is not necessary as gases get dispersed into the soil. The special feature of this type of toilet is that it has two pits instead of one. The reason being that single leach pit units are appropriate only if they can be de-sludged mechanically by a vacuum tanker, since their contents are not pathogen free. In the two pit system, the filled pit can be cleaned manually even by the householder himself because of the long period of digestion which makes it free of foul smell and safe for handling. In a single pit system, de-sludging has to be done immediately after the pit has been filled up to enable its re-use and this involves handling of fresh and undigested excreta which is a health risk and the services of scavengers would be required. If a deeper and larger single pit is provided, de-sludging operation will be difficult and there would be greater chances of pollution, especially where ground water level is high.

Sulabh Shauchalaya is also cost effective. It can be designed with different specifications and use of different type of materials without compromising the basic design principles with variable costs ranging from Rs. 500 to Rs. 10,000. It can, therefore, be afforded even by the economically weaker section. The design can also be modified suiting to the space available for construction of Sulabh Shauchalaya.

With the use of Sulabh technology dry latrines in over 7 lakh houses have been provided with Sulabh toilets and thereby 35,000 scavengers have been freed from the demeaning and dehumanising methods and practices of cleaning bucket privies and 240 towns have been made scavenging free. Sulabh International has provided the liberated scavengers with necessary training in a variety of alternative occupations/vocations that carry no social stigma. Sulabh has a work force of 35,000 persons from different disciplines viz. administrators, planners, engineers, architects, sociologists, doctors, scientists, media persons.

Pay and Use Community Toilets

This Organisation also operates and maintains more than 3,000 pay and use toilets-cum-bathroom complexes all over India. The facilities are available to men for a token charge of 50 paisa, while for women, children and the disabled it is free. There is no charge for the use of urinals. About 10 million people in India use this facility provided by the Sulabh in individual households and at community toilets with bathing and laundry facilities, popularly known as Sulabh Shauchalaya complexes.

Mobile Latrines

In many slums, there is no space for construction of community latrines with bathing facilities. Sulabh has designed mobile units mounted on wheels. The units provide toilet and bathing facilities, wash hand basins, water storage tank and collection tank for human excreta and water.

Biogas

Dr. Pathak is probably the first person in the world who has given the idea of obtaining biogas only from the excreta collected in large size public toilets used by 2,000 to 5,000 persons a day. Despite odds, he got the first such biogas plant set up at Patna in 1982 after almost six years of research. So far 63 biogas plants have been installed at the community toilet complexes and more are in the process of being set up.

Social Upgradation

Sulabh International has also taken up a programme of social upgradation of scavengers and the adoption of one family of scavenger, each by prominent social service minded citizens of the country holding prestigious position in all walks of life.

English Medium School

The Organisation has opened an English-medium public school at Palam, New Delhi for the children of scavengers who are given free modern education with thrust on vocational training, computer and office management. The school also has children from other castes and communities with the aim that age old social and class prejudices could vanish. The scavengers' children are taught through English medium so that they can complete as equal in open job market.

Technical Training

Sulabh also provides trainings to the children of scavengers for various technical courses like, carpentry, mechanic, motor driving, shorthand typing, cane work, leather work, mason tailoring and garment making, electrician, beauticians course etc on ITI pattern and syllabus. It has set up two such institutes, one at Jambhol in Maharashtra and the other at Palam, New Delhi.

Sulabh International Museum of Toilets

Sulabh has also established a Sulabh International Museum of Toilets in 1994, the first of its kind in the world. The Museum has a rare collection of facts, pictures, and objects detailing the historic evolution of toilets from 2,500 BC to date.

Dr Pathak and his Sulabh Movement

Dr . Bindeshwar Pathak, a 27 years old young social worker from Rampur Baghel, a small village in the District of Vaishali in Bihar and fresh from college took a very serious view of the ill-clad, ill-fed and emaciated men and women carrying night-soil on their heads and his own men looking upon them as if they were than apex of mankind. This changed not only his own consciousness but the future course of human history. Dr. Pathak realised that forcing this class of men to carry human excreta on their heads is an inhuman practice which becomes only an uncivilised barbarous society of ignoble savages. With the light of mission in his eyes, he launched the Sulabh Movement for the liberation of scavengers in 1970. It is twenty-six years back when he founded the Sulabh International Social Service Organisation earlier known as Sulabh Shauchalaya Sansthan, a non-profit voluntary social organisation having no capital resource of its own. It is dedicated to Gandhian ideology of emancipation of scavengers and has now acquired long and wide experience in low-cost sanitation and biogas. It also works to prevent pollution of environment and enteric diseases and promote clean living specially among the poor. In a nut-shell the Sulabh Movement is mixed package of improved technology, extensive training and social reforms different from other social reform movements in that it combines technology with idealism. The Sulabh concept and technology have been accepted as an alternative to high-cost city sewerage systems, and not only has been the status of scavengers changed in these years but a large number of them have also been liberated and settled in other professions.

Elimination of Scavenging

In the beginning the activities of Sulabh Movement were confined to Bihar alone but later Dr. Pathak realised that unless this movement is taken up at the all-India level, its impact may not be felt in other States. He, therefore, made an all-round effort to mobilise the opinion of some of the Central and State Ministers to take up a Central scheme not only for the liberation of scavengers from the obnoxious system of manual scavenging but also for their rehabilitation in other dignified professions in order to integrate them to the mainstream of the Indian social life where there is no distinction between the touchable rich and the untouchable unprivileged poor. It came to a test in March 1981 when a meeting was organised at the instance of Dr. Pathak with the then Union Minister of Works and Housing, Shri Bhisham Narayan Singh to discuss the proposal of the Government of Bihar for the liberation of scavengers. This meeting was attended by Shri Ramesh Jha, Minister of Urban Development, Bihar; Shri S.H.M. Burney, Secretary, Shri P.S. Krishnan Joint Secy. and other officers of the Ministry of Home Affairs and Secretary and other officers of the Urban Development Department of the Government of Bihar. It was decided in the meeting that a scheme may be introduced for the total elimination of scavenging from the selected towns on the whole-town approach basis. The scavengers have to be liberated from this system and given alternative employments/jobs without any loss of income and both liberation and rehabilitation have to be done simultaneously without any time gap. Since there was no scheme where this proposal could be accommodated, it was agreed that the liberation of scavengers is one of the measures for the removal of untouchability which is admissible under Sec. 15A of the Protection of Civil Rights Act, 1955 since scavengers are the victims of worst form of untouchability because of their unclean occupation and could be taken up under the Centrally Sponsored Scheme of 'Implementation of Protection of Civil Rights Act'. The State Government was asked to formulate a compreshensive scheme for the liberation of scavengers in respect of small and medium towns and this proposal should also include scheme for the rehabilitation of the liberated scavengers in alternative jobs/occupations so that they may not revert back to this inhuman occupation of scavenging. The Government of Bihar accordingly submitted the proposal to the Ministry of Home Affairs in respect of two towns viz Bihar Sharif and Purnea in March 1981 and was provided with central assistance on a matching grant basis. Thus a beginning was made from Bihar and subsequently the scheme was introduced in other States as well.

Silver Jubilee Year of Sulabh Movement

This is the silver jubilee year of the Sulabh Movement which also coincides with the 125th birth anniversary of Mahatma Gandhi. Sulabh International Social Service Organisation has built up a national network of sanitation facilities spread over 1000 towns with about 7 lakhs units in almost all States and Union Territories in India. It has been assigned to construct and maintain sanitation facilities in Nepal and Bhutan also.

One Man Forum against Age-old Practice of Scavenging

Dr. Pathak, the Founder of Sulabh International Social Service Organisation and the Sulabh Movement believes that the struggle has just begun and much has to be accomplished during the coming years. He has become a one man forum against the age-old practice of scavenging and is today the greatest champion and liberator of the scavenging class which remained oppressed and ostracised in India from time immemorial. To free scavengers from their demeaning labour and instal hand-flush latrines throughout the country, cleaning up and deodorizing city streets and providing employment and training to hundreds and potentially thousands to enable his countrymen and all who need his services to benefit from the expertise in the construction of inexpensive sanitation facilities, as the initiator of construction and maintenance of public convenience on pay and use basis, production of biogas from human excreta and its conversion into energy, are some of the major objectives and ingredients of Dr. Pathak's programme. It is in recognition of the outstanding work being done by him, he has been honoured with several national and international awards including K.P. Goenka Memorial Award for Environment for 1984, won Padma Bhushan in 1991, National Citizens Award 1991, International St. Francis prize for the Environment, 'Canticle of All Creatures' instituted by the Vatican in Rome in 1992 and the Limca Book of Records 1995. He is not content with what he has already achieved. He would rest content, he says, only when scavenging really becomes a thing of the past not only in India but other countries as well where it is still in vogue.

4

LIBERATION OF SCAVENGERS DURING FIVE YEAR PLANS

Census of Households having Toilet Facility during 1981-1991

At the time of 1991 Census* it was found that the percentage of households having toilet facility in the country was only 23.70 per cent. In other words more than 76 per cent of the people or about three-fourth of the total population in the country had no toilet facility. In the urban areas the percentage of households having toilet facility was 63.85 per cent as compared to 58.15 per cent in 1981. More than one-third of the urban households in most of the States and Union Territories did not have toilet facilities while more than half of the urban households in 177 districts did not have toilet facilities. The percentage of households having toilet facilities in the rural areas at the time of 1991 census was only 9.48 per cent.

There was an appreciable increase in the availability of toilet facility in the urban areas during 1991 census as compared to 1981. In terms of percentage the increase was registered in the States/Union Territories of Mizoram (59.92), Lakshadweep (38.03), Sikkim (24.54), Meghalaya (15.54), Kerala (13.52), Dadra & Nagar Haveli (12.31), Andhra Pradesh (10.53), Arunachal Pradesh (10.49), Goa (9.31), Nagaland (9.03), Karnataka (9.24), Punjab and Pondicherry (8.48 each), Orissa (7.39), Manipur (7.07), Tamil Nadu (6.20), Haryana (6.18), Rajasthan (5.79), Gujarat (5.60), Andamans (5.18), Maharashtra (5.08), Himachal Pradesh (4.86), Uttar Pradesh (4.48) and Bihar (3.59). The other States/Union territories which recorded only marginal increase were Chandigarh (1.24), West Bengal (1.01), Tripura (0.65) and Madhya Pradesh (0.27). It is rather surprising that the metropolitan

* Annexure IV.3 and IV.4.

city of Delhi registered a decrease in the availability of toilet facility from 68.02 per cent in 1981 to 66.64 per cent in 1991. This may be presumably due to the constant immigration of people to Delhi from other areas since the last decade.

The toilet facility was available by varying degrees to the urban households in different States and Union Territories. The lowest percentage of 49.27 was found in Orissa and the highest percentage of 96.32 in Tripura. If the States/Union Territories are arranged in ascending order according to the percentage of availability of toilet facilities based on 1991 Census, the eight North-eastern States (Arunachal Pradesh, Assam, Meghalaya, Mizoram, Manipur, Nagaland, Sikkim and Tripura) top the list (more then 81 per cent) , followed by West Bengal (78.75), Punjab (73.23), Kerala (72.66), Delhi (66.64), Uttar Pradesh (66.54), Gujarat (65.71), Maharashtra (64.45), Haryana (64.25), Karnataka (62.52), Rajasthan (62.27), Himachal Pradesh (59.98), Tamil Nadu (57.47) Bihar (56.54), Goa (55.82), Andhra Pradesh (54.60), Madhya Pradesh (53.00), and Orissa (49.27). Among the Union Territories Chandigarh (79.77) tops the list followed by Andamans (65.72), Dadra & Nagar Haveli (65.14), Lakshadweep (64.65), Pondicherry (50.02), and Daman & Diu (45.75).

Enormity of the Problem

It is estimated that nearly two billion people (about a third of World's population) are without adequate basic sanitation facilities and by the year 2000 the number may increase to 3 billion. In India the situation is not different, more than 750 million people out of 900 million population either defecate in the open or use insanitary bucket/dry privies cleaned manually or community facilities, if available. Hardly 20 per cent of the urban population have access to water flush arrangements connected to sewerage system. 14 per cent have access to water borne toilets connected to septic tanks, 33 per cent have bucket or dry privies and the remaining 33 per cent do not have any latrines in their houses. Out of nearly 4700 towns/cities there are hardly 232 towns and cities with sewerage at present. None of them, however, covers the entire municipal city area, leave alone the adjoining suburbs included in the municipal limits, in most of the urban areas even on the streets where sewers have been laid, houses have not been connected inspite of municipal laws making such connections compulsory, thus insanitation continues. In the rural areas 102 million households have no toilets facility at all.

According to the Task Force for tackling the problems of

scavengers and their rehabilitation constituted by the Planning Commission in July 1989, there were 76.4 lakhs* of dry latrines in the country, of which 54.2 lakhs were in the urban areas and 22.2 lakhs in the rural areas.

Introduction of the Scheme

The Ministry of Home Affairs while reviewing the progress of various schemes taken up by the Central and State Governments for improving the working and living conditions of the sweepers and scavengers realised that unless a comprehensive scheme giving incentive to the people is taken up, the problem of scavenging cannot be eliminated. In 1980-81 the Ministry of Home Affairs introduced a scheme of conversion of dry latrines into water pour sanitary latrines and rehabilitation of liberated scavengers and their dependents in dignified occupations in the selected towns. This scheme was taken up under the existing centrally sponsored scheme of 'Implementation of the Protection of Civil Rights Act' as one of the measures for the removal of untouchability since the scavengers were the victims of the worst form of untouchability because of their unclean occupation. Under the scheme financial assistance was provided to the State Governments on matching grant and 'whole town approach' basis. The State Governments in turn provided financial assistance to the local bodies for the conversion of dry latrines, 50 per cent of the total cost for conversion was given to the householders as subsidy and remaining 50 per cent as loan recoverable in easy instalments irrespective of income. The householder had therefore not to spend any money in the beginning on the conversion of dry latrines which was done by the local bodies itself through voluntary agencies or contractors by inviting tenders. Before a town was taken up under the scheme, a preliminary survey was carried out with a view to collecting some basic information like family composition of the scavengers, marital status, age group, educational qualification, if any etc. so as to enable the Government to chalk out suitable schemes for their rehabilitation. Since there was limitation of funds, the scheme was taken up in the selected towns only. As mentioned above, in 1980-81 the scheme was taken up in two towns of Bihar namely, Bihar Sharif and Purnea at a total cost of Rs. 1.26 crores, 50 per cent of this amount was provided by the Ministry of Home Affairs and the remaining 50 per cent by the Government of Bihar. Next in 1981-82, the scheme was extended to another eight States namely Andhra Pradesh, Assam, Kerala,

* Annexure IV.5.

Maharashtra, Rajasthan, Tamil Nadu, Tripura and Uttar Pradesh. The scheme was in operation in 16 States by the end of the Sixth Five Year Plan. The total number of towns taken up during this period was 70 where about one lakh dry-latrines were converted into pour-flush latrines and 5,000 scavengers rehabilitated in alternative employment. This was definitely a good achievement.

Absorption in the Municipalities/Local Bodies

It was found that in a large number of towns most of the displaced scavengers were observed in the municipalities/local bodies itself and nobody was rendered unemployed. The problem arose only in the case of dependents of those who were employed as private scavengers. The private scavengers were mostly women and children.

Role of Scheduled Castes Development Corporation

The Scheduled Caste Finance and Development Corporation (SCDCs) set up in various States have played a vital role in rehabilitating the unemployed scavengers by providing them with margin money loan assistance, subsidy and bank loans etc. for settling them in various trades and occupations. In some States short-term training was provided to the scavengers to enable them to settle down in various trades and occupations. The progress of rehabilitation programme was however slow due to lack of coordination between the Urban Development/ Local Self Department and the Welfare Department at the State level.

Earmarking of Funds for Rehabilitation

The Ministry of Home Affairs provided funds to the State Urban Development/Local Self Department only for the conversion of dry-latrines into water-pour sanitary latrines and not for the rehabilitation of unemployed scavengers. The work was to be done by the Department dealing with the welfare of Scheduled Castes and Scheduled Tribes at the State level out of the funds provided by the Ministry of Home Affairs to them towards share capital contribution to the SCDCs and out of SCA (Special Central Assistance). 10 per cent of the SCA was earmarked for the economic development of the vulnerable groups which included scavengers also.

Design of Latrines

Use of first pour-flush waterseal latrines with a single leaching pit for disposal of human excreta was initiated in 1943 by the all-India Institute of Hygiene and Public Health, Calcutta, and it was then known as the dug well latrine. The Research-cum-Action (RCA) Project was later taken up by the Government of India through the Health Centres of Poonamallee (Tamil Nadu), Singur (West Bengal) and Najafgarh (Delhi) in the mid-fifties. Another project was taken up at Planning, Research-cum-Action Institute (PRAI), Lucknow (UP) in 1958 and the offset double pit system was evolved. These efforts were however restricted to rural areas. In 1970 it was Sulabh International which first developed and introduced the low cost sanitation in the urban areas in a big way. Thus the offset double pit system has been introduced and found successful in different geological, hydrogeological and physical conditions.

A pour-flush latrine consists of a squatting pan of a special design (having a steep gradient at the bottom and particular depth) and a trap having 20 mm waterseal. This is designed that the human excreta of one person can be flushed with even less than 2 litres of water. The excreta is discharged into the leaching pits constructed in the house compound and where it is not possible to do so for lack of space into the pits constructed under even footpaths or streets or inside houses. The squatting pan is connected to the leaching pit through a pipe. These pits are lined with honey-comb brick work or open jointed stones so as to allow the liquid in the pits to percolate and gases to be absorbed into the soil and at the same time preventing the pit from collapsing. The sludge gets digested and settles down gradually.

The pits are used alternatively. When one pit is filled, it is stopped being used and the excreta is diverted to the second pit. The filled up pit is left unused; and in about 24 months the contents become rich organic humus, innocuous, free of pathogen and smell. When convenient, it is emptied and contents could be used as manure. It is then ready to be put back into use when the second pit becomes full in its turn.

With simple care and cleaning by the household, the pour-flush latrines can be satisfactory and hygienic sanitation system. They can be located inside the house even since water-seal prevents odour or insect nuisance.

Maintenance of Latrines

Maintenance of pour-flush latrines is very easy and simple. Day-to-day maintenance consists only of washing the latrine floor and cleaning the pan. No other maintenance cost is needed. The cost of cleaning the pits can be covered by sale of humus obtained from the pits as manure.

Cost of Conversion of Dry Latrines

The cost of conversion of a dry-latrine into pour-flush one varies from Rs. 900 to Rs. 4,500 for 5 users depending upon the area and soil. This does not include superstructure; the cost of which is to be borne by the beneficiaries themselves. Loan facilities are available for this purpose.

Reasons for Slow Progress

The progress of this scheme in the initial stage was slow on account of the following factors :

- (i) Selection of only one or two towns in each State every year under this scheme due to limitation of funds.
- (ii) Lack of willingness on the part of municipalities/local bodies to implement the scheme.
- (iii) Non-availability of good and reputed agency to execute the work. Except Sulabh International, no other NGO had dedication and expertise to undertake this type of work.
- (iv) The State Urban Development/Local Self Government Department unwilling to contribute 50 per cent of the matching share under the scheme.
- (v) Lack of coordination between the State Urban Development/ Local Self-Government Department and the Department dealing with the welfare of Scheduled Castes and Scheduled Tribes in the rehabilitation of unemployed scavengers.

Findings of Study

An evaluation of this scheme in three selected towns of Bihar and one Rajasthan in 1990 revealed that the socio-economic conditions of scavengers in the selected towns have improved in terms of income, saving and movable assets after liberation. The survey also showed that the scavengers do not suffer the kind of social and economic discrimination which they used to suffer before liberation.

Transfer of Scheme to the Newly Created Ministry of Welfare

The scheme of liberation of scavengers *inter alia* the entire work relating to the development of Scheduled Castes and Scheduled Tribes was transferred from the Ministry of Home Affairs to the newly created

Ministry of Welfare in September 1985.

Progress during Seventh Five Year Plan

An allocation of Rs. 39.07 crores was made for this scheme during the Seventh Five Year Plan. This allocation was quite meagre as compared to the magnitude of the problem. The Task Force constituted by the Planning Commission in July 1989 had estimated the number of dry latrines in the country for the year 1989 as 76 lakhs.

This programme was taken up in 226 towns from 1980-81 to 1988-89 in nineteen States. It is estimated that 3.14 lakhs of dry latrines were converted into water borne latrines and more than 10,000 scavengers liberated and rehabilitated in alternative employment by the end of March 1989 as shown in the following table:

Year	No. of towns made Scavenging free	No. of Scavengers rehabilitated
1984-85	8	1941
1985-86	2	1462
1986-87	15	2313
1987-88 } 1988-89 }	7	4342
Total	32	10058

Modification of Funding Pattern

The funding pattern of the scheme was modified from the year 1989-90 as per the recommendations of the Study Group set up under HUDCO as indicated below:

(a)	Economic Weaker Sections (EWS) (Monthly income upto Rs.1250)	45 per cent subsidy + 50 per cent loan + 5 per cent beneficiary contribution
(b)	Low Income Group (LIG) (Rs. 2651 to Rs. 4450)	25 per cent subsidy + 60 per cent loan and 15 per cent beneficiary contribution
(c)	Middle Income Group (MIG) (Rs. 2651 to Rs. 4450)	Nil subsidy +75 per cent loan and 25 per cent beneficiary contribution
(d)	Higher Income Group HIG (Above Rs. 4450)	Loan + 25 per cent beneficiary contribution

Integrated Scheme

Under the Integrated Scheme, the subsidy portion was provided by the Ministries of Welfare and Urban Development and loan by HUDCO. A Central Coordination Committee was set up under the chairmanship of the Joint Secretary, Ministry of Urban Development with the representatives of the Ministry of Welfare and HUDCO to consider proposals received from the State Governments under the scheme and decide about selection of towns, unit cost and release of subsidy etc. During 1989-90 Central assistance of Rs. 10 crores was released to the State Governments for 264 towns and Rs. 29.05 crores during 1990-91 as balance amount of Central assistance in respect of these towns.

Progress of the Scheme of Liberation of Scavengers during 1980-81 and 1990-91

It would be thus seen that the total quantum of central assistance released to the State Governments under the old scheme of 'Liberation of Scavengers' by the Ministries of Home Affairs and Welfare since its inception in 1980-81 to 1990-91 was Rs. 82.02 crores in respect of 490* towns. The number of dry latrines converted into water-borne sanitary latrines during this period was about 10 lakhs and the number of unemployed scavengers rehabilitated in alternative trades/ occupations and employment was 16,859. The number of towns made scavenging free was 40. The States of Kerala and Tripura were declared scavenger free States.

Ministry of Urban Affairs and Employment

Bifurcation of Scheme

The Planning Commission in the meeting held on the 10th September 1991 to review the programme of liberation of scavengers, decided that the Ministry of Urban Development will be incharge of conversion of dry latrines and the Ministry of Welfare will look after only the rehabilitation of scavengers. It was also decided that 45 per cent subsidy will be given to EWS category and 25 per cent to LIG. The beneficiaries have to be a 5 per cent and 15 per cent respectively of the cost as their contribution. The MIG and HIG categories will not

---

* Annexure IV.2

have the facility of subsidy. The conversion of dry latrines in their case has to be through loans and their own contribution.

The amount of loan and central subsidy is released in a synchronised manner from HUDCO directly to the local bodies or the nodal agency nominated by the State Government. The repayment period for the HUDCO loan is a maximum of seven years and the rates of interest for all categories is 10.5 per cent.

Unit Cost

The ceiling on the unit cost of Rs. 2500 for 5 users (excluding superstructure) has recently been raised for different categories as under:

(i)	5 users units	—	Rs. 3000
(ii)	10 users units	—	Rs. 4500
(iii)	15 users units	—	Rs. 5250

The additional allowance of 25 per cent over these amounts is also available to hill areas and the North-Eastern States.

As the funds earmarked for this programme are limited and there are large number of such towns with dry latrines which require priority attention, it was considered necessary to priorities the towns for implementation of this programme and accordingly it has been decided that henceforth though the present whole town approach would continue, unless all the towns having dry latrines are fully covered, no town without dry latrines would be taken up for implementation of this programme.

According to the available information, 570 towns throughout the country have been declared scavenger free. The State of Kerala, Karnataka, Gujarat and Tripura have been declared scavenger free.

The scheme received great setback after the modification of its financial pattern and the transfer from the Ministry of Welfare to the Ministry of Urban Development. The bifurcation of the scheme into two parts, each part being handled by two different Ministries, has retarded the progress of this scheme. The Ministry of Urban Development is getting the scheme implemented through HUDCO and not directly as was being done when the scheme was with the Ministry of Home Affairs/Welfare.

The implementation of the programme of Low Cost Sanitation for Liberation of Scavengers has so far been very slow as will be seen from the following table:

Year	No. of dry latrines converted	No. of new latrines constructed	No. of community latrines
1992-93	75,260	—	—
1993-94	2,76,004	2,40,021	400
1994-95	1,34,121	1,28,523	Nil

An allocation of Rs. 27.80 crores was made for this scheme during 1995-96.

Reasons of Slow Progress

The cumulative figure of achievement since the scheme was transferred to the Ministry of Urban Development has been reported as 3,90,402. The main reasons for the slow progress of this scheme can be summed up as follows:

(i) The scheme is not being implemented by the Ministry of Urban Development directly but through HUDCO. The procedure adopted by HUDCO for sanctioning loans and subsidy is too cumbersome and it takes a very long time in releasing funds. The States are therefore, rather reluctant to approach HUDCO for such financial assistance.

(ii) The allocations made for this scheme are too meagre. There is a provision of Rs. 150 crores only for this scheme during the Eighth Five Year Plan.

(iii) The funds are blocked at the State level and not released to the Municipalities/Local Bodies.

(iv) The liberated scavengers could not be provided with any employment by the municipalities/local bodies.

(v) In a number of States the scheme is being implemented departmentally or through contractors. The progress of work was satisfactory when it was entrusted to the reputed NGOs like Sulabh International etc. This agency which is in this field since 1970 and has got a band of dedicated workers, engineers and other technicians to execute the work. They take guarantee to maintain household latrines for a period of five years and public latrines for 30 years. The work done by them has been appreciated not only by the Government of India or the State Government but also by the international agencies like UNDP, World Bank etc. It is advisable to release funds to the reputed NGOs directly by the HUDCO as is being done by CAPART.

There is need to evaluate the working of this scheme through an independent agency so that the loopholes and shortcomings which come to notice are removed and remedial action taken to make it more effective. The commitment made by the Government that all the dry latrines in the country would be converted into water-borne flush latrines by 1997 and the country would be scavenging free by the end of this century would not be fulfilled unless adequate financial support is provided. The work of the conversion of dry latrines and rehabilitation of liberated scavengers should go side by side as originally envisaged under the scheme when it was started by the Ministry of Home Affairs in 1980-81.

Ministry of Environment and Forests

Under Ganga Action Plan of the Ministry of Environment and Forest there is a component of low cost sanitation for implementation in towns along the river Ganges in Uttar Pradesh, Bihar and West Bengal. The Ministry has released Central assistance to these States in respect of 36 towns for the conversion of more than 30,000 household latrines and about 15,000 community latrines within a radius of 1-2 kilometers along the river bank. This scheme has been taken up primarily from the point of view of cleaning river Ganges and not for liberating scavengers. It has no rehabilitation scheme for liberated scavengers. This scheme has very limited scope.

Ministry of Rural Areas and Employment

The problem of liberation of scavengers in the rural areas is not acute as the entire population is defeating in the fields and hardly 10 per cent of the rural population has access to sanitary facilities for the disposal of human waste. The Ministry of Rural Development has taken up low cost sanitation scheme for the construction of water-pour latrines in the rural areas but the progress is extremely slow due to certain inherent defects in the scheme.

The main strategies of the scheme are:

(i) to provide subsidy to the poorer among the households below poverty line.

(ii) to encourage other households to buy the facilities through markets, including sanitary mart.

(iii) to launch intensive campaign in selected areas and to support such campaigns with infrastructural faculties to establish individual sanitary latrines wherever possible.

(iv) to establish sanitary complexes exclusively of women wherever necessary.

(v) to encourage locally suitable and acceptable models of latrines.

(vi) to eradicate manual scavenging by converting all existing dry latrines in rural areas into low cost sanitary latrines.

Construction of Individual Household Latrines

Construction of individual households latrines by persons below poverty line is supported in cluster of 20 or more in a village. In exceptional cases like the dispersed tribal habitation/villages, the cluster may be even 10 or more.

The unit cost of construction is limited to Rs. 2500 of which 80 per cent is paid as subsidy to the selected beneficiaries below the poverty line. In case the unit cost exceeds Rs. 2,500, the additional cost is borne by the beneficiaries or specific approval of the Government of India is obtained before execution. Out of the total annual funds, a minimum of 20 per cent is earmarked for paying subsidy to the individual households from SCs and STs below the poverty line where the percentage of SC/ST population to total population below poverty line is less than 20 per cent. Where the SC/ST population below poverty line, is more than 20 per cent of the total population below poverty line, earmarking should be enhanced at least to match the percentage of SC/ST population.

Conversion of Dry Latrines

It has been decided to convert all the dry latrines (estimated to number around 24 lakhs) in the rural areas into pour-flush or other suitable type of latrines. The pattern of subsidy will be the same as in the case of individual household latrines, limiting it to households below poverty line.

Village Sanitation Complexes for Women

It has been decided by the Government to earmark upto 10 per cent of the annual funds for providing public latrines in the selected villages exclusively for women which will be constructed and maintained by the panchayats/charitable trust/NGOs.

Rural Sanitary Marts (RSMs)

United Nations International Children's Education Fund (UNICEF) have successfully demonstrated the validity of the concept of Sanitary Marts by implementing the scheme through voluntary organisations like Ramakrishna Mission in Midnapur District, West Bengal, Rural Sanitary Marts (RSMs) are outlets dealing with materials required for construction of latrines and other sanitary facilities. Besides, it also keeps the designs of various low cost sanitary facilities showing estimated cost and the list of trained masons. Rural Sanitary Marts also serve as Service Centres. UNICEF provides assistance of Rs. 50,000 for each Mart with turnover of Rs. 2 lakhs per annum for a period of two years. One time assistance upto Rs. 12,000 is provided for publicity, etc.

Where UNICEF assistance is not available for setting up RSMs, the State Governments have been empowered to set up RSMs with assistance from the Ministry of Rural Areas and Employment through reputed voluntary organisations/Pachayati Raj Institution in difficult areas where proper marketing facilities for sanitary components do not exist.

Total Sanitation of Village

Other Sanitation facilities such as drains, soakpits, solid and liquidwaste disposals, etc. are generally taken up under Jawahar Rozgar Yojna or other programme for civic amenities in the Panchayat. Where this is not feasible due to other priorities and non-availability of adequate financial resources, the facilities are taken up under the Centrally Sponsored Rural Sanitation Programme (CRSP).

Campaign for Creation of Felt-need

While government machinery for publicity is useful to some extent, a well orchestrated programme of publicity, health, education and creation of required facilities can make any change in the attitude of the people. Support of the reputed local voluntary organisations, autonomous institutions, social, political and religious organisation who carry conviction with the people are enlisted in creating the felt-need. These organisations are selected on the basis of their reputation for good and adequate infrastructure already available with them.

State Governments are also assisted to develop model villages under Integrated Rural Sanitation Programme covering faculties like sanitary latrines, conversion of dry latrines, garbage pits, soakage pits, drains, pavement of lanes, sanitary latrines in aanganwadies, schools, panchayat ghar, health centres, smokeless chullahs, cleanliness of ponds, tanks, clean surroundings around standposts, handpumps, sanitary wells and other drinking water sources with platform including washing and bathing platform, drains etc. So far the State Governments of Andhra Pradesh, Himachal Pradesh, Jummu & Kashmir and Madhya Pradesh have taken up model village schemes.

Extent of subsidy

Item	Subsidy		By contribution by
	Centre	State	user/ Panchayat
Construction of sanitary latrines and conversion of dry latrines for individual households below poverty line.	40%	40%	20%
Village Complex for women.	35%	35%	30%
Drains and other sanitation facilities (as far as possible to be met out of JRY and other funds)	25%	25%	50%
Awareness campaigns, health education demand creation etc.	Upto 10 per cent of the annual allocation.		Nil
Administrative Cost	Upto 3 per cent of the annual allocation		Nil

For Sanitary Mart, the assistance from the Central Government is limited to an interest free loan as revolving fund of Rs. 50,000 per mart for the purpose, which would be repayable at the end for three years.

Criteria for allocation of Central Assistance to States/U.Ts for implementation of CRSP and its utilisation by States

(a) 50 per cent weightage being given to incidence of poverty in States/UTs
(b) 40 per cent weightage being given to rural population.
(c) 10 per cent weightage being given to recognised hill States and hilly areas on the basis of their population.

Allocation of Funds

An outlay of Rs. 380 crores has been provided in the Central Sector and Rs. 294.23 crores in the State Sector MNP for this scheme during the Eighth Five Year Plan. The allocation made for 1994-95 was Rs. 60 crores in the Central Sector and Rs. 90.37 crores under State Sector (MNP).

Achievements—No. of Sanitary Latrines

	CRSP	MNP	TOTAL
1992-93	48,528	2,79,545	3,28,073
1993-94	1,48,878	1,82,995	3,31,873
1994-95	—	—	3,39,000

A total of 14,73,800 sanitary units have so far been constructed under this scheme. An expenditure of Rs. 135.79 crores has been incurred during 1992-93 and 1993-94 and the likely expenditure during 1995-96 was Rs. 150.37 crores.

The total rural population as estimated in 1991 census was 627.1 million. The households having sanitary facilities were about 10 per cent. This means that 112.9 million rural households are yet to be provided with basic facility of sanitation. Even at a modest cost of Rs. 2500 per unit, a gigantic investment of Rs. 28,225 crores will be required for this purpose. If a contribution of 20 per cent of the cost by the beneficiary is assumed the investment required on the part of the Government would be about Rs. 22,580 crores. Such a huge investment will be obviously beyond the means of the Centre and State Governments in the foreseeable future. In the Eighth Five Year Plan a partly sum of Rs. 674.23 crores (Rs. 380 crores in the Central Sector and Rs. 294.23 crores in the State Sector MNP) has been provided for this scheme.

The problem of rural sanitation has further been aggravated due to rapid population growth. Lack of proper sanitation in densely populated areas has become one of the most pressing problems. The lack of privacy in toilet facilities is in particular a serious problem for the women folk, old infirm and children in the rural areas. The women generally use secluded places either surrounded by trees or partly covered with crops or hidden by a raised grassy mound. The public defecation in the open is the source of inconvenience and hardship and cause of filth and disease in the rural areas. This results in the pollution of soil, water, and food which is detrimental to public health. In moist ground the larvae may be carried by the people and animals. Surface water run-off from places where people have defecated results in water pollution. In view of the health hazards created and the degradation of the environment, no open defecation should be allowed in the villages.

No reliable data is available about the exact number of dry latrines in the rural areas. The Task Force on the Liberation of Scavengers constituted by the Planning Commission as referred to earlier had estimated the number of dry latrines in the rural areas around 2.48 million* and the number of scavengers engaged in cleaning these dry latrines was about 67,200** (53,335 males and 13,865 females). We have to adopt a holistic approach (a) for conversion of dry latrines into water flush latrines, (b) for training and rehabilitation of liberated scavengers and their dependents, and (c) for prohibiting the children of scavengers from continuing this practice and sending them compulsorily to the schools with scholarships, board and lodging facilities or mid-day meal facilities.

Movement for Adequate Sanitation

Adequate sanitation is not only a pre-requisite of an acceptable standard of living for the people in the rural areas but also an effective measure to reduce health risks due to polluted environment and the transmission of excreta related diseases. There is no denying the fact that the age old habits and tradition of defecating in the open can be gradually changed and the attitude of the rural folk moulded by launching a strong movement for social reform with the help of social workers and reputed voluntary organisations. So far very little work has been done say hardly 12 per cent in this field. There is need to launch better and effective promotional and educational campaign both on social and health aspects.

* Annexure IV.5.

** Annexure II.2.

Programmes of Rural Sanitation

Besides the Central Rural Sanitation Programme which was started in 1986-87, there are a number of other programmes under which sanitary latrines are being constructed in the rural areas like Indira Awas Yojana, Employment Guarantee Programme for Rural Landless (RLEGP) and National Rural Employment Programme. There was thus, multiplicity of sanitation schemes in the rural areas. The design, cost, material, funding pattern etc. in respect of sanitary latrines under these programmes varied from scheme to scheme. There is lack of delivery system for design, material and masons etc. The involvement of beneficiaries, particularly women in these programmes was almost negligible. Since sanitation was given the lowest priority in these programmes, no tangible results could be achieved.

Nearly 20-25 per cent of the rural population are keen to install low cost pour flush latrines on their own provided the design, material and components are made available to them. It has been observed that there is a higher degree of acceptability with regard to proper sanitation facilities in areas where there is high literacy exposure to the urban way of life. It may, therefore, be easier to first cover these villages which are closer to towns where literacy rate is higher and then it may be extended gradually to other rural areas where there is demand.

Motivation of People

For the success of rural sanitation programme, people's participation is essential and it is necessary to create public awareness and education of the masses in the beneficial use of night-soil linked bio-gas plants both from the sanitation point of view and also for generation of power. The social inhibitions should be removed. Beneficiaries should be motivated to take up the programme of their own. Voluntary organisations can play a very important role in mobilising public opinion towards this programme. Motivation of people through meetings, mass media like radio, and television etc. can also be done effectively.

The experience of the past few years has amply demonstrated that unless the people—every single individual—are totally involved in the planning and implementation of this programme, very little will be achieved. The involvement has to be committed and unwavering. A sense of personal responsibility has to prevail.

There is now increased awareness of sanitation needs, constraints and possibilities. This has stimulated public and private initiatives in the rural areas to evolve and apply simple, inexpensive and socially acceptable technology and promote and support participation by community people especially women.

Role of World Bank

UNDP, whose executing agency is the World Bank, started its activities in India in the field of low-cost sanitation in 1978. They were requested to assist in preparing master plan and preliminary engineering and feasibility reports on low-cost-water-seal latrines for Assam, Bihar, Gujarat, Maharashtra, Rajasthan, Tamil Nadu and Uttar Pradesh. For this purpose, 20 towns were selected in Uttar Pradesh and 50 towns each of the other six remaining States. The reports in respect of these States, completed in 1981, covered 110 towns and generated considerable interest in the Central and State Governments. The studies, related to the conversion, existing dry latrines into low cost water-seal latrines, provision of individual low-cast water-seal latrines in houses and provision of community toilets for those households for which individual latrines could not be constructed. Similar feasibility reports were also prepared for Andhra Pradesh, Haryana, Himachal Pradesh, Jummu & Kashmir, Karnataka, Kerala, Madhya Pradesh, Orissa, Punjab, Tripura and West Bengal, Goa, Mizoram and Pondicherry. The studies done by UNDP helped the Government, launch a centrally sponsored scavenging elimination programme with 50 per cent grant to convert all the dry latrines in 33 towns in 14 States.

The low cost sanitation is also a component of the World Bank assisted Water Supply and Urban Development Projects in States like Kerala, Tamil Nadu, Uttar Pradesh, Gujarat and Madhya Pradesh. The scheme could not be taken up in all the States due to non-availability of funds.

Ministry of Welfare

As per the decision of the Planning Commission taken in the meeting held on the 10th September, 1991 the Ministry of Welfare has to look after only the rehabilitation of scavengers. The Ministry of Welfare have, therefore, formulated a comprehensive scheme for the rehabilitation of scavengers and their dependents in a manner that within a period of five years, all the scavengers and their dependents are rehabilitated in alternative occupations.

National Scheme of Liberation and Rehabilitation of Scavengers and their dependents

The National Scheme of Liberation of scavengers was launched by the Ministry of Welfare in March 1991 as a sequel to the Government's commitment to eliminate inhuman practice of manual handling of night-soil and filth by the end of the Eighth Plan period.

It is noteworthy that the scavenging has been and continues to be an hereditary occupation as a result of which by and large the children of scavengers in due course, also become scavengers. Similarly, women of the families of scavengers are also entirely engaged in scavenging.

The Task Force constituted by Planning Commission in July 1989 in its report submitted on 21st March, 1991 had estimated the number of scavengers for the year 1989 as 4,00,999 out of which 3,33,779 were estimated to be in urban areas, and 67,220 in rural areas. In the urban areas, 2,06,612 were estimated to be males and 1,27,167 females. This shows that about 83 per cent scavengers are in urban areas and 17 per cent in the rural areas. It also shows that 35 per cent are women. The estimated number of scavengers as given above is in respect of Scheduled Castes only. Taking into account the number of non-Scheduled Caste scavengers and also the growth in the number of scavengers since 1989, the number of scavengers may be roughly about 8 lakhs.

Main Component of the Scheme

The National Scheme of Liberation and Rehabilitation of Scavengers has the following components:

- (i) Time bound programme for identification of scavengers and their dependents and their aptitude for alternative trade through a survey.
- (ii) Training in identified* trades for scavengers and their dependents at the nearest local training institutes/centres of various departments of State Governments,Central Government and other semi-Government and non-Government organisations.
- (iii) Rehabilitation of scavengers in various trades and occupations by providing subsidy, margin money loan and bank loan.

---

* Annexure IV.6

The responsibility of rehabilitation of municipal scavengers in the service of local bodies would be that of local bodies themselves. The National Scheme of Liberation and Rehabilitation of Scavengers will provide rehabilitative training to private scavengers and their dependents including dependents of scavengers employed by local bodies. The scheme provides for training of scavengers and their dependents in suitably identified trades keeping in view their aptitude and local requirement and environment. The scheme also provides for incentive in terms of subsidy, margin money loan so that the scavengers and their dependents can take up alternative and dignified occupations.

Stipulation

(i) Scavenger is one who partially or wholly engaged in the obnoxious and inhuman occupation of manually removing night-soil and filth. The dependent of scavenger is one who is a member of his family or is dependent on him irrespective of the fact whether he is partially or wholly engaged in the said occupation.

(ii) The scheme would cover primarily all scavengers belonging to Scheduled Caste community. Scavengers belonging to other communities would also be covered.

(iii) It would cover scavengers in urban areas, semi-urban areas and rural areas.

(iv) There in no age limit for the purpose of rehabilitation of the scavengers. However, for the purpose of training in various trades, the age limit is 15-50 years.

(v) The duration of the training of identified scavengers and their dependents in identified suitable trades may range from one month to six months.

(vi) Training programmes would be implemented by the training institutes set up by the Government of India, State Government/ U.T. Admns. and the entire cost of the training scheme will be borne by the Central Government

(vii) The State Government would use the training centre/ infrastructure/facilities created for training in various trades by different departments of the State Governments and the Central Government and can also use the training centres/ infrastructure/facilities set up by non-Governmental organisations.

(viii) The average expenditure per trainee per month would be Rs. 500 including the stipend for the trainee which may not exceed Rs. 150 per month. This may cover the cost for providing materials, trainees fees, miscellaneous office expenses per course, stipend, trainer's honorarium, etc. The training may be provided at the nearest local training institutions so that boarding and lodging facilities are not required. The training expenditure of Rs. 500 per trainee per month would be provided to the State Scheduled Castes Finance and Development Corporations (SCDCs) for providing it to training institutions through the National Scheduled Castes and Scheduled Tribes Finance and Development Corporation (NSFDC) for conducting the training. Wherever such State Corporations do not exist, it would be provided to the State level/UT level Department incharge of Scheduled Castes welfare.

Survey and Identification of Scavengers and their Dependents

A rapid survey to identify each individual scavenger and his dependents will be undertaken and also their aptitude for specific alternative trade/occupation/job. The District Officer/District Magistrate/District Collector would be responsible for carrying out the above survey. For the urban local bodies, the survey may be carried out through the officers and employees of the urban local bodies, District Social and Welfare Officers, District Urban Development Agencies and District level SCs and STs Development corporation etc. In respect of rural areas, the survey may be carried out through the officers and employees of the Revenue Department and Rural development Department etc. The district level office of the Scheduled Castes and Scheduled Tribes corporations would be the nodal/coordinating agency for carrying out identification through suggested survey. The expenditure on the scheme for a survey for identification of scavengers may be met. if necessary, by the State Governments from the Special Central Assistance (SCA) provided to them by the Central Government for Special Component Plan (SCP). The survey was to be completed by the 31st July, 1992.

Utilisation of Existing Training Schemes

There are several training schemes of the Central Government which are being implemented through the State Governments. Such schemes are either exclusively for Scheduled Castes or they earmark a definite proportion for their coverage. These schemes include Nehru Rojgar Yojna by the Ministry of Urban Development, training schemes taken up by the Khadi and Village Industries Commission (KVIC), Small Industries Development Corporation, Craft training Schemes (ITIs) of the Ministry of Labour, Community Polytechnics by the Department of Education (Ministry of Human Resource Development) and TRYSEM by the Department of Rural Development. The National Scheme of Liberation and Rehabilitation of Scavengers will have to take full advantage of the existing training schemes of the Central and State Governments. However, they are required to design special training schemes for scavengers so that they are suitably tailored to meet the requirements of these people. The focus will have to be on creating/upgrading skill for self-employment. It is not necessary to prescribe any minimum qualification for providing training to the scavengers.

Rehabilitation Programme

This scheme has the following components:

(i) Time-bound programme for identification of scavengers and their dependents and their aptitude for alternative trade through a survey.

(ii) Immediate need for identification of trades and preparation of shelf of projects.

(iii) Training with stipend in identified trades for identified scavengers and their dependents at identified training centres of various departments of the State Government, Central Government and other semi-Government and non-Governmental organisations.

(iv) Phasing of coverage under the scheme.

(v) Funding pattern including provision of subsidy, margin money loan and bank loan.

(vi) Monitoring and evaluation.

(vii) Effective coordination with such departments and other organisations at District, State and National level which are already implementing such poverty alleviation programmes and training programmes and other programmes which aim at providing viable self-employment.

(viii) drawing up special training programmes for scavengers as may be necessary by different training institutes.

Financial Assistance for Self Employment

This scheme provides for funding of projects costing upto Rs. 50,000 per beneficiary and also for margin money to the extent of 15 per cent of the project cost at 4 per cent rate of interest. The project costing above Rs. 50,000 would need prior approval of the Ministry of Welfare.

For a project costing Rs. 50,000 the break-up would be Rs. 10,000 subsidy, Rs. 7,500 margin money loan from the State Scheduled Castes Development Corporation and Rs. 32,500 loan from the banks.

Under the scheme, subsidy would be 50 per cent of the project cost upto Rs. 15,000 and 50 per cent subsidy in projects costing from Rs. 15,001 to Rs. 50,000 with maximum ceiling of Rs. 10,000. Thus for financial assistance the maximum project cost would be Rs. 50,000 with 50 per cent subsidy with a maximum ceiling of Rs. 10,000 and 15 per cent as margin money. The State Scheduled Castes Corporations would stand surety to the bank loan and the assets to be created would be mortgaged to the Corporations. Presuming that the average cost per rehabilitation project for one scavenger/dependent would be Rs. 20,000 the total requirement of funds for 4 lakh scavengers and their dependents would be Rs. 800 crores.

Keeping in view the subsidy and margin money loan pattern stated above, the break-up of funds required would be as follows:

(a)	Subsidy	Rs.400 crores
(b)	Margin money loan from the share capital of SCs Dev. Corporations. Central Share — Rs. 58.80 crores. State Share — Rs. 61.20 crores.	Rs. 120 crores
(c)	Bank loans	Rs. 280 crores
		Rs. 800 crores

The total Central share in the cost of rehabilitation including the cost of training is estimated to be Rs. 563.795 crores or say Rs. 564 crores and the corresponding State share towards margin money loan is estimated to be Rs.61 crores.

The expenditure on training and subsidy is borne fully by the Government of India on 100 per cent basis. The expenditure on margin money loan will be shared between the Central and the State Governments in the ratio of 49:51 as prescribed under the Centrally Sponsored Scheme of Share capital cortribution to the State Scheduled Castes Development Corporations.

Implementing Agencies

SCDCs were designated as nodal agency for the implementation of the scheme at the grass-root level. The National Scheduled Castes and Scheduled Tribes Finance Development Corporation (NSFDC) has been given the responsibility of day-to-day implementation and monitoring of the scheme as well as the task of coordination and collaboration with the SCDCs in implementing the scheme. The District Collectors/Deputy Commissioners were made responsible for overall implementation of the scheme.

Pattern of Assistance for Rehabilitation and Training

The rehabilitation of scavengers in alternative trades/occupations and employment is achieved by providing a package of financial assistance upto Rs. 50,000 per beneficiary which includes a maximum subsidy of Rs. 10,000 or 50 per cent of the total unit cost, margin money loan at concessional 4 per cent rate of interest constituting 15 per cent of the project cost and balance as loan from the commercial banks.

The Central Government reimburses training cost @ Rs. 500 per trainee per month which also includes Rs. 150 per trainee per month as stipend. The amount of stipend is too inadequate since persons engaged in scavenging earn more than Rs. 750 per month. There is need to raise the amount of stipend atleast upto Rs. 750. The duration of training which has been fixed at maximum of 6 months is also too short for courses of technical nature and should be raised to atleast one year for technical courses where training is imparted by I.T.Is and Polytechnics. The need for flexibility in designing training courses has also to be examined.

Identification of Scavengers through Survey

The survey of scavengers which was to be completed by June 1992 is still not complete in the States of Bihar, Orissa, West Bengal and Gujarat. The survey has so far identified 7,36,114 persons. The surveys conducted also suffer from deficiencies and there have been complaints that many localities, families have been left out.

Progress

The scheme has been in operation for the last five years but the progress* achieved during this period was extremely dismal barring a

* Annexure IV.6

few exceptions as is evident from the following table:

Year	Central Assistance released (Rs. in crores)	Number of scavengers	
		Trained	Rehabilitated
1991-92	50.50	—	—
1992-93	60.73	16,288	19,007
1993-94	70.97	13,266	40820
1994-95	73.00	24,568*	63746*
1995-96	90.00	50,000*	1,00,000*
Total	345.20	1,04,122	2,23,573

* Provisional

Central Assistance blocked at State level and not released to SCDCs

It is rather surprising that though the Central Government had released a huge amount of Rs. 345.20 crores to the State Governments for the rehabilitation of scavengers and their dependents from 1991-92 to 1995-96, an amount of Rs. 88.86 crores only could be released by the State Governments to the SCDCs which are actually implementing the scheme and the remaining amount is blocked at the State level and thereby earning interest on it.

Poor Progress

In so far as the achievement of physical targets in concerned, it is far from satisfactory. As against the total number of 4 lakhs of scavengers, 1,50,000 have been rehabilitated during the last five years which constitutes only 37.5 per cent of the total number of scavengers.

Central Monitoring Committee

The Ministry of Welfare has constituted a Central Monitoring Committee under the chairmanship of Welfare Minister to coordinate, review and monitor the implementation of the programmes under the scheme of 'Liberation and Rehabilitation of Scavengers and their dependents'. Though the committee was constituted in January, 1993, it has met only once.

It is envisaged in the scheme that there would be a State-level* Monitoring Committee under the chairmanship of the Chief Secretary with representatives from the Departments of Urban Development,

* Annexure IV.7

Local Bodies, Industries/Small Scale Industries, Rural Development, Labour, Education, Technical Education, Institutional Finance, Planning, Finance and two representatives each from scavengers community and lead banks which is required to meet once in three months. There will be a similar committee at the District level under the District Magistrate/Deputy Commissioner and also at the Local-body level.

Some of the State Governments have constituted the committees at the State and District* level but a large number of them have not set up such committees. The committees which have so far been set up are on paper only excepting two or three State Governments.

National Commission for Safai Karamcharis

The Government of India constituted a National Commission for Safai Karamcharis under the National Commission for Safai Karamcharis Act, 1993 in August 1994 under the chairmanship of Shri Mangi Lal Arya with six other members including a Vice-chairman Shri J.P.D. Salappa. The tenure of the Commission is upto 31st March, 1997. The main function of the Commission is to recommend to the Central Government specific programmes of action towards elimination of inequalities in status, facilities and opportunities for Safai Karamcharis under a time-bound action plan; study and evaluate the implementation of the programmes and schemes relating to social and economic rehabilitation of Safai Karamcharis and make recommendations to the Central Government and State Governments for better coordination and implementation of such programmes and schemes; investigate specific grievances and take *suo moto* notice of matters relating to non-implementation of :

(i) Programmes or schemes in respect of any group of Safai Karamcharis;
(ii) decisions, guidelines or instructions, aimed at mitigating the hardships of Safai Karamcharis
(iii) measures for the social and economic upliftment of Safai Karamcharis, and take up such matters with the concerned authorities or with the Central or State Governments and make periodical reports to the Central and State Governments on any matter concerning Safai Karamcharis, taking into account any difficulties or disabilities being encountered by them.

* Annexure IV.8

N.C.S.K. not empowered to study and evaluate Low Cost Sanitation Schemes

The Commission has not been given powers to study and evaluate the implementation of the programme of Low Cost Sanitation and Liberation of Scavengers handled by the Ministry of Urban Development and HUDCO. Unless the Commission is given necessary power to study and monitor all the schemes with which the Safai Karamcharis are directly or indirectly related, the problems of these people cannot be solved in right earnest and there will not be full justification in the setting up of such a Commission.

5

PROSPECTS AND RETROSPECTS

Even fifty years after independence, the inhuman and degrading practice of manual scavenging of night-soil which is still prevalent till today in certain parts of the country is an affront to human dignity and a social evil. The pace of implementation of schemes to eliminate this curse has been tardy and slow in almost all States in the country. We have not been able to get rid of the blot of not being able to secure inalienable right of all scavengers to live with dignity.

The Government of India did launch a scheme some years back with great fanfare to rehabilitate scavengers within a definite timeframe, but the scheme came a cropper. Scavenger families still lead a socially degrading life in the urban areas. The crucial issue of right to live with dignity has now been taken up by the National Human Rights Commission which has urged all the State Governments to take urgent steps to eliminate this practice and to rehabilitate scavengers within a definite time-frame.

Slow Progress

The scheme of "Liberation of Scavengers" which was started by the Ministry of Home Affairs in 1980-81 under PCR Act with the twin objectives of conversion of existing dry latrines into pour flush latrines on the whole town approach basis in the selected small and medium towns; and rehabilitation of liberated scavengers in alternative dignified occupations suffered great setback after it was transferred to the Ministry of Urban Development and HUDCO with the modified financial pattern. The 50 per cent subsidy which was available to all householders irrespective of income, has been restricted to Economically Weaker Sections to the extent of 45 per cent of the unit cost and in the case of Low Income Groups to the extent of 25 per cent and balance as loan and beneficiary's contribution. No subsidy is now available to other income groups like MIG/HIG. The procedure followed by HUDCO in sanctioning subsidy and loan is too cumbersome and municipalities/local bodies are rather reluctant to approach HUDCO for this purpose. This is evident from the releases made during the last few years. The pace of progress of the scheme has thus been considerably slowed down and in some States it has virtually come to a standstill as hardly 4 to 5 lakhs of dry latrines have been converted into water borne flush latrines in the country during the last five years. The liberated scavengers could not be provided with any employment by the municipalities/local bodies.

Lack of Coordination

The Ministry of Welfare which is responsible for the rehabilitation of scavengers is implementing the scheme in isolation without knowing whether the scavengers have been liberated from doing manual scavenging or still engaged in this obnoxious system of scavenging. There is no coordination between the Ministries of Welfare and Urban Development either at the Central or State levels.

Allocation during Eighth Plan

An outlay of Rs. 464 crores was provided in the Eighth Plan for the rehabilitation of 4 lakhs of scavengers as against the actual requirement of Rs. 905 crores—Rs. 105 crores for training and Rs. 800 crores for rehabilitation in various trades and occupations (Rs. 400 crores as subsidy, Rs. 120 crores margin money loan at 4 per cent interest and Rs. 280 crores bank loan). The above calculation is based on the average expenditure of Rs. 20,000 per beneficiary, which includes 50 per cent of the total cost of a project or a maximum of Rs. 10,000 as subsidy, Rs. 3000 (15%) margin Money loan and balance or Rs. 7000 as bank loan. The maximum cost of a project is Rs. 50,000.

Rehabilitation in Haphazard Manner

The scheme has been in operation for the last five years but the progress has so far been very slow. The rehabilitation programme has been implemented in a very haphazard manner. The statistics furnished by the Government shows that though 60,000 scavengers were trained during 1991-92 to 1995-96 1,50,000 scavengers were rehabilitated during the same period but hardly 60,000 scavengers could actually be rehabilitated. Large number of beneficiaries were provided assistance for setting up trades/occupations without any training. The projects chosen for rehabilitation are also more or less a mirror image of projects under the IRDP programme. The project cost per beneficiary was reported to be as low as Rs. 10,000 as against the maximum project cost of Rs. 50,000 envisaged under the scheme in many States but with effect from 1.4.96 the average cost has been increased to Rs. 50,000. The coverage of family for rehabilitation would also include group projects. It is an admitted fact that the rehabilitation of the beneficiaries in professions without adequate training would not enable the beneficiaries to switch over to such trades and the possibility of then going back to the old profession would remain strong in view of the low returns from such projects.

Reluctance of Commercial Banks to Finance Projects

Another contributory factor for the slow progress of the scheme is the reluctance of the commercial banks to finance projects under the scheme. Inspite of the Reserve Bank of India's derectives to the commercial banks, instances have come to notice where the banks have rejected loan application from beneficiaries on flimsy ground, or they have insisted on collaterals. It is the duty of the district officials to co-ordinate with the lead bank of the district and ensure that banks advance loans under the priority sector lending programme.

Transfer of Funds to SCDCs

The other reason for the unsatisfactory progress of the scheme was delay in transfer of central assistance from the State Government to the Scheduled Castes Finance and Development Corporation (SCDCs). The State Government have been delaying the release of central assistance under the scheme to SCDCs which are the implementing agencies. Many of the States have not released funds upto 3 years, keeping the central funds in civil deposits. The Government of India had decided that from 1.4.96 Central assistance will be released directly to the SCDCs and not through the State Governments.

Weakness of Training Scheme

So far as the training part of the scheme is concerned, a meagre amount of Rs. 500 per month per trainee has been provided which includes Rs. 350 as training expenses and Rs. 150 towards stipend. The amount of stipend is quite inadequate and is not able to attract sufficient number of trainees since each scavenger earns more than Rs. 750 per month. The duration of training is only one to six months which is too short for courses of technical nature. The duration of training courses will, therefore, have to be suitably increased for courses of technical nature like motor mechanic, computer, stenography etc. where training is imparted by ITIs and polytechnics etc. There is also need for adopting flexibility in designing training courses. The decision to utilise only existing training centres run by the Central and State Governments like Nehru Rozgar Yojana of the Ministry of Urban Development, training centres of KVIC, Craftsmen Training Schemes (ITIs) of Labour Ministry, Community Polytechnics of the Ministry of Human Resource Development and TRYSEM of the Ministry of Rural Development etc., also need reconsideration since there are a number of areas where no training facilities are at present available. The Government of India has now decided to adopt the TRYSEM norms for the purpose of training under this programme with effect from 1st April, 1996. The scavengers and their wards may not like to go to distant places for training without any hostel facilities. It would be, therefore, desirable to entrust such training programmes to the reputed non-Government social service organisation which are committed to the cause of scavengers like Sulabh International Social Service Organisation which has got branches all over the country and is already running such a training centre at Palam, New Delhi. Sulabh has already trained more than 3,000 liberated scavengers in various trades and occupations in Bihar. There is another centre at Jambhol, Thana District, Maharashtra run by the Sulabh International Social Service Organisation in collaboration with the State Government. The centre provides training to all the liberated scavengers in a number of trades which enabled them to get employment or settle down in the trades/occupations of their choice.

No Time Limit for Abolishing Scavenging

The Parliament in the Budget Session of 1993 passed the 'Employment of Manual Scavengers and Construction of Dry-latrines (Prohibition) Bill, 1993' to abolish scavenging and put a ban on the construction of dry latrines and engage in or employ for or permit to be engaged in or employed for any other person for manually carrying human excreta. This was no doubt, a historic decision but no time limit has been given by which the system of manual scavenging would be totally abolished and thus the very purpose of the Bill has been defeated. The states of Andhra Pradesh, Goa, Karnataka, Maharashtra, Tripura and West Bengal had already adopted this Act by passing a resolution in their respective Assemblies. The state Governments of Bihar and Delhi have also followed suit from this year.

Conclusion

It was precisely four years ago when the Government took a decision to eliminate manual scavenging from the country within a period of five years. This process which has two parts, namely, conversion of all the existing dry latrines into pour flush latrines and construction of new sanitary latrines for those who defecate in the open, and rehabilitation of liberated scavengers in alternative jobs and occupation after giving them short training, wherever necessary, is at a crucial stage. We have many more miles to go and many more bridges to cross. If for financial or any other reasons, the scheme is bogged down, it will have serious repercussions on the reforms process in India. It is,therefore, necessary to evolve some kind of broad concensus among the State Governments to ensure that the country is made scavenging free atleast by 2006 and to ensure that the last scavenger is properly rehabilitated in alternative employment by that time.

The term 'Bhangi' considered to be derogatory is used for those manually cleaning and carrying human excreta: Ironically it is mandatory under the Constitution (Scheduled Castes) Order, 1950 to use the term 'Bhangi' in order to get the benefits as Scheduled Castes. And if the term is not used in the Scheduled Caste Certificate, the applicant will be denied of the benefits. The Constitution (Scheduled Castes) Order, 1950 and other subsequent orders therefore, need modification in order to substitute the term 'Bhangi' with some dignified name such as 'Balmiki' etc.

ANNEXURE I.1

NAMES OF CASTES, SUB-CASTES/TRIBES, SYNONYMS AND SURNAMES IN VARIOUS STATES WHOSE TRADITIONAL OCCUPATION IS/OR WAS SCAVENGING AS GIVEN IN THE ETHNOGRAPHIC RECORDS

The Tribes and Castes of the North-Western India by W. Crooke (1896)

1. Bhangi
2. Chuhra, Chura or Chuhara
3. Mehtar
4. Halalkhor
5. Khakrob
6. Baharwala
7. Musalli
8. Lalbegi, Shaikhri, Dumri, Heli
9. Balmiki, Valmika
10. Dhanuk
11. Hela
12. Patharphor
13. Hanri or Hari
14. Bansphor
15. Dhe
16. Rawat
17. Shaikhs
18. Baniwal
19. Bilpurwar
20. Tank
21. Gahlot
22. Kholi
23. Gagra
24. Sarohi
25. Chandaliya

26. Sirsawal
27. Siriyar
28. Turaiha

Names connected with some tribe or occupational and well known caste	*Names from their place of origin*
1. Bagri	1. Antarbedi
2. Bais	2. Duab
3. Baiswar	3. Balkhariya
4. Balakchamariya	4. Banaudh
5. Bargiyar	5. Baranwar
6. Barwar	6. Bhojpuri Rawat
7. Bhadauriya	7. Ghazipuri Rawat
8. Bisensob	8. Jamalpuriya
9. Bundeliya	9. Jamunapari
10. Chamariya	10. Janakpuri
11. Chandela	11. Jaunpuri
12. Chauhan	12. Kanhpuriya
13. Chhipi	13. Katheriya
14. Dhelphor	14. Manglauri
15. Gadariya	15. Manikpuri
16. Jadou	16. Mainpuri
17. Jadubansi	17. Mathuriya
18. Jaiswar	18. Mehtaranpuri
19. Jogiya	19. Mukundpuri
20. Kachhwaha	20. Moultani
21. Kayasthbansi	21. Nanakpuri
22. Kinwar	22. Sayyidpuri
23. Sakarwar	23. Sarwariya
24. Tank	24. Ujjainwal or Ujjainpuri
25. Thakur Bais	
26. Turkiya	

Aligarh	1. Chutele	**Kanpur**	Basor
	2. Kalawata		Domar
	3. Kharautiya	**Fatehpur**	Supa Bhagat
	4. Kothiya		
	5. Kausikiya		
	6. Mathuriya	**Allahabad**	Bilkhariya
Mathura	Soda	**Jhansi**	Domar

District	No.	Name
Mainpuri		Pattharwar
Etah	1.	Churele
	2.	Katheriya
	3.	Mathuriya
	4.	Pattargoti
Bareilly	1.	Bargujar
	2.	Dankmardan
	3.	Janghare
	4.	Katheriya
	5.	Rajuriya
	6.	Khariya or Kharaha
	7.	Dalwariya
	8.	Tank Mardan
	9.	Singha
Bijnor		Gangwati
Moradabad	1.	Barchi
	2.	Bargujar
	3.	Bhumiyan
	4.	Deswali
	5.	Multani
	6.	Rajauriya
Shahjahanpur		Katheriya
Mirzapur	1.	Halalkhora
	2.	Lal Begi
	3.	Rawat
	4.	Domar
	5.	Hinduaiya
	6.	Kirtiya
	7.	Hela, Malwar
Ghazipur		Rawat
Basti	1.	Audhiyar
	2.	Desi
	3.	Dom
Lucknow	1.	Bansphor
	2.	Hela
	3.	Rawat
	4.	Hari
	5.	Dhanuk
	6.	Lal Begi
	7.	Dom
	8.	Shaikh or Shaikhra
	9.	Chuhra
Unao		Turaiha
Sultanpur		Dom
Benares	1.	Lal Begi
	2.	Shaikh
	3.	Mehtar
	4.	Hela
	5.	Rawat
	6.	Chauhan or Chuhan
	7.	Kharaha
	8.	Patthara
Kheri	1.	Mehtar
	2.	Bhangi
	3.	Lal Begi
	4.	Chaudhari
	5.	Rangreta
	6.	Hathile
	7.	Rawat
	8.	Domra
	9.	Dhubae
	10.	Bansphor

Important local sub-castes

Dehradun The Badlan and Nanakshahi

Saharanpur
1. Barlang
2. Chanahiya
3. Machal
4. Tank

Muzaffarpur
1. Bhilaur
2. Deswal
3. Gahlot
4. Soda

Bulandshahr
1. Bachanwar
2. Baiswair
3. Bhadauriya
4. Bhagwatiya
5. Bhokar
6. Chandaliya
7. Chauhan
8. Chauhela
9. Chunar
10. Duakauliya
11. Goranthiya
12. Jangphare
13. Jasnuleali
14. Nauratan
15. Nirbani
16. Panwari
17. Phulpanwar
18. Rathi
19. Rolapal
20. Shaikhawat
21. Tarkhariya
22. Turkiya
23. Ujjainpuriya
24. Ujjainwal

According to Crooke the total number of scavenging castes in the North-Western India in 1891 Census was 1406 (Hindu castes—1359 and Muslim castes—47)

Castes and Tribes of North-Western Provinces (Punjab) by Ibbetson (1916)

1. Chuhra
2. Dhanak
3. Khatik
4. Bhangi
5. Kutanas
6. Mazbi
7. Rangreta
8. Musalli
9. Dumna
10. Sahotra
11. Gil
12. Bhatti
13. Khokhar

14. Mattu
15. Kharu
16. Kaliyana
17. Ladhar
18. Sindhu
19. Chhapriband
20. Untwal
21. Kandabari
22. Hansi
23. Khosar
24. Borat
25. Dhariwal

A Glossary of the Tribes and Castes of the Punjab and North-West Frontier Province (H.A. Rose—1919)

1. Chuhra
2. Mazbi or Mazhabi
3. Musalli
4. Kurtana or Kotana
5. Dindar
6. Dindar-Kharob
7. Bhangi
8. Deswali
9. Sotarwala
10. Bagri
11. Jangalke
12. Balmiki
13. Lal Begi
14. Lohat
15. Bohat
16. Sarwan
17. Lute
18. Jhae, Dhae or Sahi
19. Tengre
20. Goriye
21. Kandara
22. Laddar
23. Khokar
24. Khonje
25. Kaliane
26. Ratti
27. Mathi
28. Burt
29. Mome
30. Hauns
31. Chapriban
32. Ghussur
33. Balhim
34. Labante
35. Nahir
36. Dum
37. Mirasi
38. Machchi
39. Jhiwar
40. Changar
41. Ghunjur Beg
42. Ail Maluk
43. Pir Chhota
44. Bala

People of India, Haryana, Vol. XXIII (K.S. Singh—1994)

1. Balmiki, Valmiki
2. Chuhra
3. Bhangi
4. Lalbegi
5. Khakrobe
6. Chaudhary
7. Jamadar

Gotras

8. Pihwal
9. Dugalgach
10. Tank Saudde
11. Baid
12. Marholia
13. Tanwar
14. Chauhan
15. Pathak
16. Rajput
17. Chauhan
18. Jangalia
19. Panche
20. Kunjinwal
21. Gaichand
22. Bagdi
23. Lohat
24. Luchare
25. Lohaye
26. Chandalia
27. Athwal
28. Khekhra
29. Bhooranda
30. Mazbi

The Tribes and Castes of Bombay by R.E. Ethoven (1920)

1. Bhangis or Bhangias
2. Halalkhors
3. Olgana
4. Barvashia
5. Metariya
6. Jamphoda

7. Mela
8. Bhasods
9. Chajgadis
10. Helas
11. Lalbegis
12. Makhiyars
13. Sheikhs
14. Bhuimali

Surnames

1. Chohan
2. Chudasama
3. Dafda
4. Jethva
5. Makwana
6. Solanki
7. Vaghela
8. Vadhiya
9. Dhedva
10. Kumbhar
11. Maru
12. Purbiya

Tribes and Castes of the Central Provinces of India by R.V. Russel and Hiralal (1916)

1. Mehtar
2. Bhangi
3. Hari
4. Dom or Dumar
5. Lalbegi
6. Ghasia
7. Mahar
8. Rawat
9. Shaikh
10. Hela
11. Makhia
12. Nornolia
13. Chuhra Mehtar
14. Dhanuk
15. Bansphor
16. Shaikh Mehtar
17. Makhia
18. Nornolia

The Castes of Marwar by Munshi Hardyal Singh (1894)

1. Bhangis - Loharis and Multanis
2. Lalbegis
3. Khakrobs
4. Halalkhor
5. Mehtar
6. Lakhan
7. Goyal
8. Jangra
9. Tamboli
10. Jajoran
11. Danoria
12. Ugach
13. Chhawa
14. Sarsar
15. Bund
16. Gund
17. Rel
18. Teje
19. Zinda
20. Pamat
21. Dhikiya
22. Chavaria
23. Chandanaliya
24. Channaliya
25. Gujrati
26. Mattu
27. Bhiman
28. Thagaria
29. Dhanwar
30. Gudwar
31. Hatwar
32. Daora
33. Adiwal
34. Kilana
35. Dharu
36. Umarwal
37. Changra
38. Chahan
39. Gamal Guch
40. Sonkhat
41. Mirga

42. Karelia
43. Pandat
44. Bharya
45. Channariya

The Tribes & Castes of Bengal by H.H. Risley (1891)

1. Hari
2. Mihtar
3. Har-Santan
4. Bhuinmali
5. Bara-Bhagiya or Kaora-paik
6. Madhya-Bangali
7. Maghaya
8. Karaiya
9. Purandwar
10. Khore or Khoriya
11. Sinli
12. Madhya-bhageya or Madhaukul
13. Bangali

Cosmogony of Caste and Social Mobility in Assam by Bimal J. Dev and Dilip K. Lahiri (1984)

1. Bansphor
2. Lalbegi
3. Hari
4. Maghaia
5. Halkhor
6. Dom, Doom
7. Ghashi
8. Uria
9. Goal
10. Tanti
11. Hela
12. Kurmi
13. Turi
14. Koiree
15. Goar
16. Kahar
17. Bagti
18. Sammar
19. Kol

20. Sheck
21. Chuna Naga
22. Kabui Naga
23. Shekra
24. Raut
25. Dumar
26. Kurmai

People of India, Goa, Vol. XXI (K.S. Singh—1994)

1. Bhangui
2. Waghele
3. Parmar
4. Multani
5. Goher
6. Chavan
7. Nandrekar
8. Chatribin
9. Chavaria

The Castes and Tribes of H.E.H., the Nizam's Dominions by Syed Siraj-ul Hassan (1920)

1. Lalbegis
2. Hele
3. Malkane
4. Chhichhade
5. Dumar
6. Chhajgade
7. Bhadeye Dassan
8. Makhyar
9. Phatrod
10. Sanakat
11. Kanderas
12. Suraswal
13. Gaikwad
14. Sarwan
15. Kanderia

ANNEXURE I.2

DISTRIBUTION OF SCHEDULED CASTE WORKERS ENGAGED IN SCAVENGING (1961 CENSUS)

	Sch. Caste workers engaged in scavenging	Percentage to total SC workers	Rural	Urban	Total workers in scavenging and percentage of Sch. Caste Workers	
India	3,86,725	1.28	1,71,303 (44.30%)	2,15,423 (55.70%)	8,02,400	(48.20)
1. Andhra Pradesh	10,845	0.35	2,838	8,007	43,883	(24.71)
2. Assam	1,080	0.35	204	876	7,018	(15.38)
3. Bihar	17,027	0.53	3,681	13,346	37,340	(45.49)
4. Gujarat	19,341	3.32	3,980	15,361	35,111	(55.08)
5. Jammu & Kashmir	292	0.25	202	90	5,230	(05.58)
6. Kerala	1501	0.22	148	1,353	15,742	(09.53)
7. Madhya Pradesh	32,030	1.34	11,797	20,233	58,720	(54.54)
8. Madras	23,060	0.71	4,826	18,232	38,518	(59.86)
9. Maharashtra	19,587	1.73	2,445	17,139	72,535	(27.00)

(Contd.)

ANNEXURE I.2 *(contd.)*

	Sch. Caste workers engaged in scavenging	Percentage to total SC workers	Rural	Urban	Total workers in scavenging and percentage of Sch. Caste Workers	
10. Mysore	13,970	0.86	1,447	12,523	30,174	(46.29)
11. Orissa	7,845	0.58	2,992	4,494	6,573	—
12. Punjab	8,819	4.74	42,947	25,872	1,04,454	(65.88)
13. Rajasthan	39,271	2.36	24,428	14,843	50,822	(77.25)
14. Uttar Pradesh	1,09,804	1.61	65,634	44,130	2,19,406	(50.04)
15. West Bengal	8,548	0.36	1,755	6,793	48,943	(17.46)
16. Delhi	12,804	10.88	1,486	11,310	24,480	(52.30)
17. Himachal Pradesh	973	0.43	350	623	1076	(90.42)
18. Tripura	219	0.52	60	159	297	(73.73)
19. Dadra & Nagar Haveli	2	0.39	2	-	13	(15.38)
20. Manipur	-	-	-	-	217	
21. Pondicherry	71	0.23	42	29	796	(08.91)
22. Nagaland	-	-	-	-	237	

ANNEXURE II.1

POPULATION OF SWEEPERS AND SCAVENGERS IN 1931 CENSUS

Sl. No.	States	Bhangi	Chuhra	Halalkhor	Hari	Lalbegi	Mehtar	Paki	Remarks
	All India 19,57,460	7,97,599	7,21,981	23,165	3,18,830	5145	88,723	2017*	1038678 M 918782 F
1.	Ajmer-Merwara	6,980	-	-	-	-	-	-	
2.	Bombay	94,950	-	-	-	-	-	-	
3.	United Provinces	5,03,608	-	-	-	-	-	-	
4.	Baroda	31,026	-	-	-	-	-	-	
5.	Gwalior	28,527	-	-	-	-	-	-	
6.	Rajputana	99,083	-	-	-	-	-	-	
7.	Western India States	33,425	-	-	-	-	-	-	
8.	Baluchistan	-	5,213	-	-	-	-	-	
9.	Delhi	-	21,138	-	-	-	-	-	
10.	North Western Forntier Province	-	9,285	-	-	-	-	-	
11.	Punjab	-	6,81,359	-	-	-	-	-	

Contd.

ANNEXURE II.1 *(contd.)*

Sl. No.	States	Bhangi	Chuhra	Halalkhor	Hari	Lalbegi	Mehtar	Paki	Remarks
	All India 9,57,460	7,97,599	7,21,981	23,165	3,18,830	5145	88,723	2017	
12.	Jammu & Kashmir	-	4,986	-	-	-	-	-	
13.	Bengal	-	-	876	1,46,735	4,965	23,281	-	
14.	Bihar & Orissa	-	-	22,289	1,15,613	157	-	-	
15.	Assam	-	-	14,348	-	-	2,364	-	
16.	Madras	-	-	-	42,134	-	-	2,017	
17.	Sikkim	-	-	-	-	23	3	-	
18.	Central Province & Berar	-	-	-	-	32,182	-	-	
19.	Central India	-	-	-	-	30,893	-	-	

ANNEXURE II.2

STATEWISE ESTIMATES OF THE NO. OF SCAVENGERS IN 1989

States/UTs		1961 Census Code 931		Persons Engaged in Scavenging Code 931		%age of Scavengers		1981 Census Code		Estd.no of Scavengers in 1981		Estd. no. of Scavengers in 1989	
		Male	Female	Male	Female	Male	Female	Male	Female	Male	Female	Male	Female
1	2	3	4	5	6	7	8	9	10	11	12	13	14
Andhra	R	6946	3690	1739	1099	25.04	29.78	12858	6057	3220	1804	3812	2086
Pradesh	U	20671	12576	4693	3314	22.70	26.35	25718	18602	5838	4902	6296	5537
	T	27617	16266	6432	4413	23.29	27.13	38576	24659	9058	6706	10108	7623
Assam	R	2327	158	181	23	7.78	14.56	-	-	-	-	-	-
	U	3766	767	753	123	20.00	16.03	-	-	-	-	-	-
	T	6093	925	934	146	15.32	15.78	-	-	-	-	-	-
Bihar	R	8875	5789	2461	1220	27.73	21.07	8168	1705	2265	359	2187	15
	U	14615	8151	8019	5327	54.87	65.35	20996	10280	11521	6718	12922	7274
	T	23490	13940	10480	6547	47.61	46.97	29164	11985	13786	7077	15109	7289
Gujarat	R	5179	2260	2826	1154	54.57	51.06	7202	1986	3930	1014	4372	958
	U	18252	9420	8813	6548	48.29	69.51	24946	14872	12046	10338	13339	11854
	T	23431	11680	11639	7702	49.67	65.94	32148	16858	15976	11352	17711	12812

Source : Report of the Task Force on the Problems of Scavengers constituted by the Planning Commission (1990).

ANNEXURE II.2 *(contd.)*

1	2	3	4	5	6	7	8	9	10	11	12	13	14
Haryana	R	-	-	-	-	-	-	10859	2694	-	-	-	-
	U	-	-	-	-	-	-	11345	4073	-	-	-	-
	T	-	-	-	-	-	-	22204	6767	-	-	-	-
Himachal	R	361	65	289	61	80.05	93.85	1658	494	1327	464	1742	625
Pradesh	U	406	244	382	241	94.08	98.77	2084	914	1961	903	2593	1168
	T	767	309	671	302	87.48	97.73	3742	1408	3288	1367	4335	1793
Jammu &	R	1945	80	197	5	10.13	6.25	1608	157	163	10	149	12
Kashmir	U	2732	473	69	21	2.53	4.44	4043	972	102	43	115	52
	T	4677	553	266	26	5.69	4.70	5651	1129	265	53	264	64
Karnataka	R	3359	1361	905	542	26.94	39.82	5852	1271	1577	506	1846	492
(Mysore)	U	18835	6619	8298	4225	44.06	63.83	20278	8250	8934	5266	9188	5682
	T	22194	7980	9203	4767	41.47	59.74	26130	9521	10511	5772	11034	6174
Kerala	R	5303	3295	115	33	2.17	1.00	5068	6748	110	67	108	181
	U	3431	3713	815	538	23.75	14.49	3610	6750	857	978	874	1154
	T	8734	7008	930	571	10.65	8.15	8678	13498	967	1045	982	1335
Madhya	R	11069	12624	5701	6096	51.50	48.29	9971	5217	5135	2519	4909	1088
Pradesh	U	20853	14174	11204	9029	53.73	63.70	30765	17769	16530	11319	18660	12235
	T	31922	26798	16905	15125	52.96	56.44	40736	22986	21665	13838	23569	13323
Maharashtra	R	10535	3727	1369	1076	12.99	28.87	14505	4265	1884	1231	2090	1293
	U	40548	17725	9262	7877	22.84	44.44	65051	28593	14858	12707	17096	14639
	T	51083	21452	10631	8953	20.81	41.74	79556	32858	16742	13938	19186	15932

ANNEXURE II.2 *(contd.)*

1	2	3	4	5	6	7	8	9	10	11	12	13	14
Meghalaya	R	-	-	-	-	-	-	120	102	-	-	-	-
	U	-	-	-	-	-	-	504	463	-	-	-	-
	T	-	-	-	-	-	-	624	565	-	-	-	-
Manipur	R	124	7	-	-	-	-	169	37	-	-	-	-
	U	86	-	-	-	-	-	206	64	-	-	-	-
	T	210	7	-	-	-	-	375	101	-	-	-	-
Mizoram	R	-	-	-	-	-	-	51	5	-	-	-	-
	U	-	-	-	-	-	-	160	113	-	-	-	-
	T	-	-	-	-	-	-	211	118	-	-	-	-
Nagaland	R	96	1	-	-	-	-	162	40	-	-	-	-
	U	140	-	-	-	-	-	189	64	-	-	-	-
	T	236	1	-	-	-	-	351	104	-	-	-	-
Orissa	R	1740	660	1281	1710	73.62	259.09	4929	2368	3629	6135	4568	7905
	U	2457	1716	2161	2333	87.95	135.96	9363	5994	8235	8149	10665	10475
	T	4197	2376	3442	4043	82.01	170.16	14292	8362	11864	14284	15233	18380
Punjab	R	27762	40426	15582	27365	56.13	67.69	5035	2920	2826	1977	-2276	-8178
	U	23422	12844	15122	10750	64.56	83.70	18100	6551	11685	5483	10310	3376
	T	51184	53270	30704	38115	59.99	71.55	23135	9471	14511	7460	8034	-4802
Rajasthan	R	12948	11807	11321	13107	87.43	111.01	13616	8056	11904	8943	12137	7277
	U	16964	9113	7665	7178	45.18	78.77	25148	11523	11362	9077	12841	9837
	T	29912	20920	18986	20285	63.47	96.96	38764	19579	23266	18020	24978	17114

(Contd.)

ANNEXURE II.2 *(contd.)*

1	2	3	4	5	6	7	8	9	10	11	12	13	14
Sikkim	R	32	3	-	-	-	-	128	14	-	-	-	-
	U	28	1	-	-	-	-	265	34	-	-	-	-
	T	60	4	-	-	-	-	393	48	-	-	-	-
Tamil Nadu	R	5876	1615	3965	861	67.48	53.31	11925	3928	8047	2094	9680	2587
(Madras)	U	21492	9535	11633	6601	54.13	69.23	32606	17478	17650	12100	20057	14300
	T	27368	11150	15598	7462	56.99	66.92	44531	21406	25697	14194	29737	16887
Tripura	R	94	14	50	10	53.19	71.43	157	64	84	46	99	60
	U	132	57	105	54	79.55	94.74	254	171	202	162	241	205
	T	226	71	155	64	68.58	90.14	411	235	286	208	340	265
Uttar	R	79873	67295	33373	32301	41.78	48.00	30424	12713	12711	6102	4446	-4378
Pradesh	U	50813	21425	28784	15346	56.65	71.63	69661	24295	39463	17403	43735	18226
	T	130686	88720	62157	47647	47.56	53.70	100085	37008	52174	23505	48181	13848
West Bengal	R	6681	1117	1438	,317	21.52	28.38	6072	1866	1307	530	155	615
	U	34334	6811	4723	2070	13.76	30.39	37268	8299	5128	2522	5290	2703
	T	41015	7928	6161	2387	15.02	30.11	43340	10165	6435	3052	6545	3318
Arunachal	R	-	-	-	-	-	-	163	30	-	-	-	-
Pradesh	U	-	-	-	-	-	-	99	35	-	-	-	-
	T	-	-	-	-	-	-	262	65	-	-	-	-
Andaman &	R	31	2	-	-	-	-	126	11	-	-	-	-
Nicobar	U	121	43	-	-	-	-	307	71	-	-	-	-
Islands	T	152	45	-	-	-	-	433	82	-	-	-	-

ANNEXURE II.2 *(contd.)*

1	2	3	4	5	6	7	8	9	10	11	12	13	14
Chandigarh	R	-	-	-	-	-	-	112	17	-	-	-	-
	U	-	-	-	-	-	-	3655	1041	-	-	-	-
	T	-	-	-	-	-	-	3767	1058	-	-	-	-
Delhi	R	1349	827	843	643	62.49	77.75	2922	1362	1826	1059	2219	1225
	U	17624	4680	9063	2255	51.42	48.18	35963	13663	18492	6583	22264	8314
	T	18973	5507	9906	2898	52.21	52.62	38885	15026	20318	7642	24483	9539
Dadra and	R	10	3	1	1	10.00	33.33	-	-	-	-	-	-
Nagar Haveli	U	-	-	-	-	N.A	N.A.	-	-	-	-	-	-
	T	10	3	1	1	10.00	33.33	-	-	-	-	-	-
Goa, Daman	R	19	21	-	-	-	-	232	258	-	-	-	-
& Diu	U	99	107	-	-	-	-	690	653	-	-	-	-
	T	118	128	-	-	-	-	922	911	-	-	-	-
Pondicherry	R	381	243	40	2	10.50	0.82	190	209	20	2	12	2
	U	77	95	11	18	14.29	18.95	654	539	9.3	102	126	136
	T	458	338	51	20	11.94	5.92	844	748	113	104	138	138
North-East	R	177	17	-	-	-	-	-	-	-	-	-	-
Frontier	U	-	-	-	-	-	-	-	-	-	-	-	-
Agency	T	177	17	-	-	-	-	-	-	-	-	-	-
Laccadive	R	14	-	-	-	-	-	-	-	-	-	-	-
Minicoy	U	-	-	-	-	-	-	-	-	-	-	-	-
Amindivi Islands	T	14	-	-	-	-	-	-	-	-	-	-	-

(Contd.)

ANNEXURE II.2 *(contd.)*

1	2	3	4	5	6	7	8	9	10	11	12	13	14
All India	R	193196	157107	83677	87626	43.33	55.77	154282	64594	61965	34862	53355	13865
	U	311898	140288	131575	83818	42.19	59.77	443928	202127	18495	114755	206612	127167
	T	505004	297395	215252	171474	42.62	57.66	598210	266721	246922	149617	259967	141032

Source : 1. Census of 1961, Code 931, Census of India, 1961 Vol. I Part II -B (ii) General Economic Tables.

2. Scavenging, 1961: Census of India, 1961 Vol. I Part V-A (i) Special tables for Scheduled Castes.

3. Code 541+542 Census of India, 1981 Part - III-B and (iv) Table B-18 General Eco. Tables.

ANNEXURE III.1

RECOMMENDATIONS OF THE REPORT OF BARVE COMMITTEE (1949)

NO.F. 13 -118-/54-LSG
GOVERNMENT OF INDIA
MINISTRY OF HEALTH
New Delhi. Dated 24th September, 1955

From

Shri A.V. Venkatasubban, M.A.
Deputy Secretary to the Government of India.

To

All Part 'A', 'B' and 'C' State Governments

SUBJECT : Report of the Scavengers' Living Conditions Enquiry Committee

Sir,

I am directed to say that on the recommendation of the Backward Classes Board, the Government of Bombay appointed a Committee in 1949 with Shri V.N. Barve as Chairman to study and enquire into the living conditions of the scavengers in the State of Bombay and to suggest ways and means to improve their present conditions of work and fix their minimum wages. The said Committee submitted its report to the Government of Bombay in 1952. The Government of India presume that the Government of Madras etc., have got copies of this report which was published in 1952.

2. The Government of India feel that though the recommendations made by the Scavengers' Living Conditions Enquiry Committee, specifically relate to the Local Bodies in the State of Bombay, they are really capable of wider application and can with profit, be adopted by all the States. The Government of India, therefore, suggest that the Government of Madras etc., may take steps to implement as many recommendations of this Committee.

A summary of the important recommendations made by the Scavengers' Living Conditions Enquiry Committee is enclosed. The receipt of this letter may kindly be acknowledged and the Government of India informed of the action taken in the matter in due course.

Yours faithfully,
Sd/- A.V. Venkatasubban
Deputy Secretary

Summary of the important recommendations of the Scavengers Living Conditions Enquiry Committee

The Committee was appointed some time in 1949 and submitted its report in 1952. It relates only to living conditions of scavengers in State of Bombay but some of the points raised therein are of common interest to all State Governments.

(1) It may be noted with interest that latrines with receptacles which are generally removed by scavengers came into vogue only after establishment of municipalities during the British regime.

(2) The caste of Bhangi as such was confined only to certain Provinces like Gujarat, Maharashtra, Karnataka, etc. It was of a very late origin. Ancestors of these Bhangis were just field labourers of a low caste but never did the work of scavenging. Some of these people took to the dirty work of cleansing the latrines for the sake of profit. Slowly this developed into a monopoly. The stage was reached when the Bhangis wanted to exploit this monopoly and a sort of customary right was thus developed. By force of habit the Bhangi lost his self-respect to such an extent that he did not consider the dirty work of cleansing latrines as a curse from which he should endeavour to extricate himself.

(3) The Committee are strongly of the opinion that the Gharaki system or the systen of customary right of scavenging should be abolished both in the interest of the Bhangis and also in the interests of public sanitation. The latter reason has arisen because Bhangis who had customary rights had the upperhand over the residents of town and villages because they would do the cleansing work only as and when they pleased and at the same time they would not allow other scavengers to enter into their sphere of work to carry on the work in case of strikes etc. This led to a high degree of insanitation in many cases.

(4) The Committee felt that suitable measures should be taken to awaken the sanitary conscience and civic sense of the public in the matter of abolition of the process of removal of night-soil by hand by the Bhangis, and to raise culture of the Bhangis caste and made them fit to take their due share in the public life of the society.

(5) The Committee has pointed out that in certain Municipal Acts, there is provision to regulate the construction,

maintenance and control of water-closets, privies, latrines, etc., but there is no provision for the scavenging of public latrines. This has prevented the municipalities from breaking the so-called customary rights of Bhangis and from improving the methods of scavenging in order to avoid the hand removal.

(6) Certain Municipal Acts, for instance in Punjab and Delhi, actually include provision for the preservation of the customary rights of scavenging private latrines. These provisions should be abrogated. (In this connection attention may be invited to our proposed legislation to amend the Punjab Municipal Act in order to do away with customary right in the Delhi City).

(7) The living conditions of scavengers are very poor. Provision of housing of Bhangis or payment of house rent to such workers who are not provided with free quarters does not find a place in all Municipal Acts. Wherever such provision has not been made the Act should be amended suitably.

(8) Provision of water for drinking or washing in the case of scavengers is very inadequate. Even where storage tanks and water taps are provided it has been found that the water supply is non-existent after certain hours with the result that the scavengers are put to a lot of suffering.

(9) The area where scavengers reside do not have any latrines. These have to be provided.

(10) The quarters and localities for scavengers are often situated in unhealthy places in the vicinity of open drain or dump grounds. This should be remedied. In view of the fact that the breaking of monopoly of customary right of scavengers will take some time, the question of housing accommodation should be tackled immediately by local bodies.

(11) There should also be provision for adequate lighting of streets and open spaces, for a playground, for a congregational centre, for a reading room, etc. The local bodies should be asked to pay attention to such things in the case of scavengers also. Unless these amenities are provided slum clearance cannot really be a success.

(12) House rent should be paid to all scavengers who have neither a house of their own or who have not been provided free accommodation by local bodies.

(13) The dirtiest feature of scavenging work is the process of hand-removal of night-soil from the ordinary latrines, in which the night-soil drops down in a chamber below, where usually a basket or some other receptacle such as a pan, a bucket, etc.,

is placed. The Committee have recommended that such receptacle latrines should be eliminated forthwith and substituted by a type of latrines which will not involve hand-removal.

(14) Local bodies whenever they have occasion to construct new latrines should definitely avoid constructing the usual pan latrines and should construct septic-tanks or other suitable substitutes. Local bodies should also refuse permission to new house owners to construct such receptacle latrines.

(15) The Committee have also made recommendations as to the various types of latrines that can be brought into use in order to avoid this hand-removal. They have stated that the practice of carrying the night-soil baskets and drums on the head from the latrines to the night-soil cart or lorry should be prohibited. The cart or the lorry should pass along the road as the work of cleansing latrines proceeds and should keep pace with it. A wheel-barrow to which a drum of suitable size to carry night-soil with cover should be fixed and the same should be pushed by hand up to the night-soil cart or lorry, otherwise the night-soil may be carried in buckets with cover. In the case of cess-pools, a suction pump fitted to a lorry should be utilised. If this is not possible the wheel-barrow with a drum or any other device may be adopted.

(16) It may be interesting to note that the Committee have suggested that the night-soil should be used for the production of gas for street lighting, fuel, etc. There are gas plants already manufactured in the country which can be used for this purpose. Public latrines should be so constructed that the night-soil etc., should enter special tanks built for the plant from which gas will be produced and after the gas has been yielded the remaining stuff is thrown out at the other end of the tank and would still retain its manurial value. The Committee have recommended that the potentialities of such gas plants for use in village panchayats should be examined. (In this connection it may be recalled that Mr. Thirin Narayanan of the Malaria Institute of India has already evolved a gas plant for the production of gas from cow-dung and other refuse).

(17) The Committee have also recommended that five hours of scavenging work should be the maximum for each scavenger. The practice of giving off-days should also be adopted.

(18) Local bodies should provide suitable uniforms or clothing to the scavengers in their employ.

(19) A campaign must be planned by the Public Health Authorities of States, with the cooperation of local bodies and Social workers for the purpose of securing improvements in the living condition of the scavengers and generally for the uplift of scavengers as members of society.

(20) The Committee have made recommendations in regard to the minimum wages which should be prescribed for scavengers They have proposed a scale of Rs. 25—1/2—30 in the case of district municipalities with a selection grade of 15 per cent with a further time scale of Rs. 5 spread over 10 years and in the case of Notified Area Committee and Gram Panchayats at a flat rate of Rs. 20 per month. The same scale of allowances as is allowed to Class IV employees of Government should be given to the scavengers also. In certain municipalities an allowance called an unclean work allowance is also given. This allowance should be treated as part of the wages of scavengers.

(21) If as a result of improvement in the conditions of work of the scavengers as mentioned above, some of them are thrown out of employment, it should be the duty of the local bodies to see that such persons are absorbed in other suitable kind of employment under their control.

(22) Government should appoint an Assistant to each Director of local authorities in a State in order to secure the implementation of the Committee's recommendations.

(23) Local bodies, Gram Panchayats and other local authorities should include in their annual reports a special chapter giving details about the living conditions and the conditions of work of scavengers and sweepers in their employ and the improvements effected therein during the year under review.

(24) Cooperative Credit Societies should be organised with the help of local bodies for the use of the servants of local bodies including scavengers.

(25) Lastly, the Committee have recommended that the sanitary conscience, civic sense and the sense of social justice of the people in general should be aroused by impressing upon them the importance of public sanitation and improved sanitary arrangements in their own dwellings.

ANNEXURE III.2

REPORT OF THE SCAVENGING CONDITIONS ENQUIRY COMMITTEE
(MALKANI COMMITTEE—1960)
Chapter Two

Cleaning of Latrines

The Receptacle

1. The use of receptacles in private and public latrines should be made compulsory.
2. The receptacle should be made of iron sheets or M.S. plates of suitable gauge to last at least a year or so and should be provided with a handle on one side and a slanting side on the other.
3. The non-use of receptacles in latrines should be made an offence and the defaulters dealt with properly.
4. The Municipal bye-laws should be so framed as to penalise the use of receptacle other than the prescribed size and make.
5. The house-owners left to themselves would not go in for the receptacles even though it is made obligatory and penalty imposed for failure to provide the receptacle. The receptacles should, therefore, be supplied periodically, at least once a year or more if the first is broken or damaged, by the local body directly or through licensed shops on a no-profit-no-loss basis, and if necessary at subsidised rates.
6. It should be the duty of the scavenger and the area sanitary inspector to report to the local body about a receptacle which might have broken or damaged.
7. The local body should recover the cost of receptacle along with the scavenging tax where such a tax is in force or with other taxes at the time of their realisation.
8. The floor of latrines is often 'kutcha'. It should be cemented and slanting outward.
9. The receptacle chamber should not be too wide to allow lateral or forward movement of the receptacle. It should be placed in a fixed spot so as to ensure that the aperture in the squatting plate is just over the centre of the receptacle. Fixed place can be ensured by making a slight depression of the size of the

receptacle in the floor with smooth slanting sidings so that the receptacle may get fitted into the cavity and water if any accumulated can be wiped out conveniently. The depression should not be deeper than half an inch.

10. Providing of receptacles in public latrines should be the responsibility of the local bodies.
11. Bigger receptacles so as to cater to the needs of the place should be provided in public latrines of big *chawls katras* or houses having large number of people.
12. The number of public latrines should conform to the demand for them in the area on the basis of their usage. It is desirable to prescribe a standard proportion to the population of the area and conform to it.
13. The structure of latrines where no receptacle can be placed at present should be got changed so as to make provision for a receptacle. This would not be costly and would not require much change.
14. As far as possible, all public buildings like offices, schools, hospitals etc.,should have flush out latrines failing which septic-tank latrines or aqua-privies be provided. But where this is not possible, it should be ensured that standardized receptacles are provided. Hotel owners and owners of places where large number of people stay should also follow this practice.
15. Wherever new buildings, colonies or townships are built by Government or with Government aid or by local bodies and in all industrial housing programmes, the latrines must be of sanitary types.

Night-soil, urine and ablution water to remain separate

16. It would be better if the latrines are so modified as to ensure keeping night-soil and urine or ablution water separate. This can be done by having a slope in front of the seat so that urine is not mixed with night-soil. An additional step may be provided side by side so that the squatter shifts to the third step for purpose of washing.

The Scraper

17. It should be obligatory to use a standardized scraper which should be supplied by the local bodies to their employees and not left to the scavengers themselves.

18. In case of private scavengers, the local bodies should supply the standardized scraper directly or through licensed dealers at no-profit-no-loss basis and it should be made obligatory to use such a scraper.
19. The scraper should be of iron sheet or M.S. plates with a long handle at one end and a semi-curve at the other end. The sheet or plate should be of a suitable gauge so that it may not break while collecting the night-soil. The corners of the plate should be rounded.
20. It should be the duty of the scavengers and the area sanitary inspector to ensure that broken or damaged scrapers are replaced immediately.

The Rubber Gloves

21. Efforts should be made to encourage the use of rubber gloves, particularly at the pail-depots and dumping grounds.

The Head Load

22. Carrying of night-soil in basket, *tasla* and drum etc. as head load should be prohibited by law and made an offence.
23. For stopping carrying of night-soil as head load, persuasive methods should also be adopted and they can be more effective than legislation.

The Bucket

24. In narrow lanes or lanes which are unpaved or have ups and downs, wheel barrows cannot ply conveniently and cannot be taken right up to the latrine seat. For such cases, there is no alternative but to introduce buckets. The bucket when full can be carried to the wheel-barrow standing nearby.
25. In narrow lanes or roads of varying altitude, means like *Kawar* also known as *Kawath, Kawad* or *Bengi* may be used.
26. The local bodies should supply small buckets of not more than 2 to 3 gallons capacity, painted with coal-tar to its employees.
27. The lid of the bucket should be fixed with hinges as otherwise this is either thrown away or not used. The lid should be collapsible.
28. Private scavengers should also be compelled to use small

covered buckets which can be supplied to them by the local bodies on no-profit-no-loss basis.

29. The broken or damaged buckets should be replaced immediately.

The Cess-pools

30. Collection of night-soil in cess-pools is most insanitary and the inhuman ill use of scavengers engaged in cleaning them should be stopped.
31. The contents of cess-pools must not be carried as head load.
32. Cess-pools maintained by the local bodies particularly those attached to public latrines, should not be allowed to continue. All the pail-depots where night-soil is collected in big drums should be closed down.
33. Abolition of cess-pools in private houses is likely to take some time. As long as these are in existence, the local bodies must introduce mechanical devices for cleaning them periodically. There should be a suction system.
34. In no case the contents of cess-pools should be removed by hand but only through mechanical devices.
35. Wherever feasible, lorries like scampers should be introduced for sucking the contents of the cess-pools.
36. Lorries fitted with suction pump can be made use of.
37. Corporations or big municipalities or such local bodies as may afford, should think seriously in terms of introducing vacuum cars/lorries etc., as used in Japan of cleansing cess-pools.

Washing of Latrines

38. It should be ensured that latrines are washed properly and regularly.
39. The scavengers should be given access to latrines.

Chapter Three

Transportation of Night-soil

40. After collection, night-soil is taken in basket, *tasla* drum etc., either as head load or waist load to the pail-depot, carriage van or the place of disposal. This is most inhuman. The entire

process of carrying night-soil as head load or waist load which is the root cause of all the troubles, has to be revolutionised.

41. For putting an end to the practice of carrying night-soil as head load or waist load, wheel barrows should be introduced.
42. The bullock or *bhensa* carts used for transporting night-soil are most insanitary and deserve to be abolished at once.
43. The man-drawn carts for carrying night-soil are an insult to human dignity and must be abolished.
44. There are various other types of carts drawn by bullocks which are all primitive. These need immediate replacement.
45. The open trucks are as bad as bullock carts. These should be discouraged.
46. The local bodies should introduce improved types of night-soil vans, tankers and trailers etc.
47. Some local bodies collect night-soil in overhead tanks or big drums which have to be poured into night-soil vans for transportation. In some places, night-soil is churned and collected in pools before transportation. Such storage tanks and churning operations must not be allowed to continue.
48. In places of over-head tanks or pools used of collecting night-soil, the scientific system of providing trailers or night-soil tankers should be encouraged.
49. The night-soil barrels used for collection purposes should be placed at higher levels so that night-soil can be poured out right into the night-soil lorry tanker, or better still there should be a suction system.

Chapter Four

Wheel-Barrows

50. The crude types of wheel-barrows already in use in some States should be discarded.
51. Some States are using drum type of wheel-barrows of 15 to 40 gallons capacity. Such wheel-barrows need replacement with better designs.
52. The design of wheel-barrow as given in Figure No. 49 of the committee's report is considered to be most useful, the introduction of which would facilitate the work of scavengers. This design could, however, be modified or improved in the light of local conditions.
53. The frame of wheel-barrow should be all steel one with solid

rubber tyre wheels with ball bearings. There should be two pails in the wheel-barrow. The pails should not be of more than 5 gallons capacity each and the combined capacity of both the pails should not exceed the capacity of 8 gallons when full. The lids of the pails should be collapsible and fixed with hinges to the body of the pails. The bottom of the pails should be such that night-soil does not stick.

54. Even women scavengers would not resist the use of such light wheel-barrows. At some places, women have already taken up to wheel-barrows.
55. The design of wheel-barrow proposed would not normally cost more than Rs.100 to Rs.120 if made locally or Rs. 120 to Rs. 140 if imported from any other State.
56. The manufacture of the proposed type of wheel barrow on a large scale for requirements throughout a State could be centralised to some place for distribution to all local bodies in which case the cost of wheel-barrow would definitely come further down and the design will conform to the required pattern uniformly.
57. A few of the municipalities are using wheel-barrows with 4 pails. Such wheel-barrow costs Rs. 175 to Rs. 200. These wheel-barrows are quite heavy and the scavengers using them complained of their weight. Such wheel-barrows can be used by men only any not women.
58. Wherever necessary, some administrative adjustments may be carried out. Men may do the pushing and women the cleaning. The wheel-barrows may be kept moving from house to house along with the women workers.
59. For places where wheel-barrows cannot ply conveniently. *Kawath* also known as *Kawar, Kawad* or *Benghi* may be used. Tow pails each of 4 to 5 gallons capacity hanging at either end of a bamboo pole at a distance of about 6 feet or so and may be carried on shoulder like the milk cans which are found convenient in such cases. An iron hook at either end may also be fitted to the pole.
60. The main difficulty in popularising the introduction of wheel barrows is the passive attitude of scavengers. This requires tactful handling. The local authorities should make a beginning with a few intelligent scavengers and with the assistance of non-official organisations and social workers, persuade scavengers to use wheel-barrows.
61. To begin with, only a few wheel-barrows may be put into use

on an experimental basis and on their becoming popular, more wheel-barrows may be pushed in.

62. The Public Health Officers can by persuasion and by amending the bye-laws etc., achieve substantial measure of success.
63. Even the local bodies which have plans for underground drainage system, must take up the scheme of wheel-barrows without further delay.
64. The number of wheel-barrows can be reduced with the extension of the underground drainage line.
65. For transporting night-soil, the bullock carts, *bhensa* carts, donkeys gangs etc., which are used at present should be replaced with tractors and trailers or such other mechanical vehicles.
66. In places where the trailers may not be convenient to ply, the local authorities may make use of three wheelers The intake capacity of the three-wheelers is quite good and these can ply in narrow streets easily.
67. The trailers must be placed at places within easy approach of scavengers and where these are needed more.

The entire process of dealing with night-soil from latrine to places of disposal would thus involve—

(a) collecting of night-soil from receptacle under the latrine-seat with a proper scraper into the pail of a wheel barrow or a bucket provided with a collapsible lid where wheel barrow cannot ply:
(b) emptying of bucket into the pail of wheel-barrow;
(c) carrying of wheel-barrow to the trailer or pail-depot wherever it is in use and emptying the pails into the trailers etc.; and
(d) drawing of trailer by the tractor to the dumping ground or the place of disposal.

Financial Aspects

68. For the administration of various schemes, the main burden will have to be borne by the local bodies concerned with or without the assistance from the State Governments. They will have to make necessary adjustments and provisions in their bye-laws, rules etc.
69. The State Governments will have to ensure that their local

bodies agree to adopt the measures uniformly throughout the State.

70. For purchase of wheel-barrows, the local bodies find it difficult to find out adequate financial resources. The best course would be for Government of India to meet the entire cost towards the purchase of wheel-barrows; failing which:

(a) the poorest local bodies should be given 100 per cent assistance—75 per cent to be given by the Central Government and 25 per cent by the State Governments—the entire amount to be treated as grant;
(b) for medium local bodies, the Governments grant can be 75 per cent —50 per cent to be given by the Central Government and 25 per cent by the State Governments and the local body to contribute only the remaining 25 per cent from its own resources;
(c) for the bigger municipalities and corporations etc., the Government of India's assistance should be 50 per cent, the rest to be borne by the local body alone or by the local body and the State Government concerned together.

The Union Territories being under the administrative control of the Central Government, the entire expenditure on schemes is borne by the Central Government. Being so, the Union Administrations are supposed to take a lead in the abolition of carrying night-soil as head loads.

71. In all cases where the local bodies concerned need loans for meeting their share of expenditure, the State Governments should help the local bodies by advancing loans to them.

72. For purchase of other mechanical equipments like tractors and trailers or vans etc., the State Governments should find the fund which may be advanced to the local bodies as grants wherever possible or otherwise in the form of loans. If possible, necessary funds may also be found from the agricultural programmes for converting night-soil into manure. If the Ministry of Health provides any funds for such purpose that may be made use of.

73. The local bodies should earmark a fixed percentage out of their annual budgets for the improvement of scavenging conditions and the living conditions of scavengers.

74. Wherever wheel barrows are supplied to private scavengers, it should be the responsibility of the local body to watch the

proper use and maintenance of the wheel barrows through the area sanitary inspectors.

Long-handled Brooms

75. Long-handled broom with a small *Khurpi* (flat scraper) fixed at the other end for scraping purposes would be found very convenient.

Chapter Five

Implements

76. Implements' allowance—Instead of giving implements' allowance to scavengers, local bodies should supply all the necessary and proper implements to their employees to carry out their duties.
77. Besides the implements suggested by the Committee, the Public Health Officers should apply their minds to the problem and think about the continuous improvements of implements.
78. The local bodies by adopting necessary bye-laws should see that even private scavengers are equipped with necessary implements for use.

Chapter Six

Night-Soil Depots

79. The number of pail-depots must be increased according to the needs. The pail-depots should be constructed at places where it is convenient and within easy reach of the scavengers.
80. In towns where due to over-crowding there is no place for constructing additional pail-depots, there trailers can be used more conveniently. These can be stationed at suitable places and taken away with the help of a tractor at periodical intervals.
81. The existing conditions of the night-soil pail-depots and the scavengers working there are extremely unsatisfactory and deserve the immediate attention of the local bodies.
82. The work at pail-depots must be carried out in hygienic conditions and efforts should be made to ensure that the scavengers avoid bodily touch with night-soil and rubbish.
83. Cess-pools for storing night-soil before transportation should not be allowed to continue.

84. The workers at night-soil-depots must be provided with gum-boots, gloves, aprons, water-proof headgears, long spades and rakes etc., for pulling out cloth pieces etc., to avoid clogging.
85. The structure of such depots where perforce, the scavengers have to wade through night-soil must be altered. Provision of trailers can minimise the difficulty to a great extent.
86. The depots may be so constructed as to allow the trailers to clear the load in the mouth of the depot itself.
87. In case of dry pail-depots, platforms of proper height with a sloping siding at either ends just like railway paltforms must be constructed so that the wheel-barrows can be brought with ease over the platform and poured into the standing trailer.
88. At every pail-depot, there must be an attendant. It should be his duty to handle the load at the depot.

Amenities at Pail-Depots

89. At every pail-depot arrangements for the supply of sufficient water for washing the vehicles, wheel-barrows, drums, the depot etc., clothes must be provided.
90. There may be separate bath-rooms for men and women. A set for each can be sufficient in most of the cases. The number may be increased if necessary.
91. Wherever possible, arrangements for changing clothes before and after the work should be provided.

Disinfection Stations

92. Corporations, big municipalities and such other local bodies as can afford, should establish disinfection stations where clothes of the scavengers may be given steam treatment. The local bodies may extend the benefit to others such as labourers, hospitals etc., also.
93. Wherever the scavengers do not find it convenient to keep the wheel-barrows etc., at the places of their residences, arrangements for keeping the wheel-barrows at the pail depots may be made. Their cleanliness can also be ensured.

Persons Working in Gutters

94. For cleaning underground sewers, use should be made of mechanical devices already in vogue in big towns. Safety of

the life of persons working in gutters should be ensured and such implements, apparels, masks, etc., as may be considered necessary be provided to them.

Chapter Seven

Dumping Grounds

95. Dumping grounds should be as far away from the towns as possible. Where towns have extended, dumping grounds should be shifted further beyond.
96. The size of the dumping ground should be sufficient so as to meet the requirements.
97. All the night-soil and refuse of the town should be brought to the dumping ground. This would increase the income from the sale of manure.
98. To collect all the night-soil and refuse from the town, proper vehicles for transport should be supplied and vigilance maintained to see that no night-soil is wasted by the scavengers or members of the public.
99. As at present, most of the dumping grounds are running at a loss. With the exception of a few isolated cases, the position need not be so tight and there is no reason why the entire business should not be turned into a profit making one and at least the cost of maintenance be realised.
100. To suppress the stink and nuisance of flies at dumping grounds, use of the Swingfog or the Gammaxine Powder or any such other useful chemical must be made by the local bodies.

Workers at Dumping Grounds

101. Long gum-boots (up to the knee) are an absolute necessity for workers at the dumping grounds. They must be provided with gloves as well. An apron would also protect the clothes. Necessary implements for use should be given to the workers. They should be supplied with working uniforms and headgears.
102. The extra allowance of a rupee or so given by some of the local bodies to workers at dumping grounds does not meet the requirements.
103. At all dumping grounds, arrangements for the supply of sufficient water for washing vans, pails etc., and for scavengers to take their baths and wash their clothes must be made.

104. Construction of a few bath rooms would be very useful.
105. For cleaning the vans etc., overhead water tanks and hoses can be more useful. The ground underneath should preferably be cemented.
106. At places where water supply is scarce, the local authorities must endeavour to supply as much water as possible. Digging of a well at the site may solve the problem to some extent.
107. In rocky areas, a well is not a practical solution. There the local bodies should make arrangements for bringing water in vans and storing in overhead tanks etc.
108. Somehow or the other, water must be made available at the dumping grounds.
109. Supply of soap at the dumping grounds and also at pail-depots is essential.
110. Arrangements for liquid soap or soap cakes may be made wherever possible.
111 Shelters may be provided at the dumping grounds under which the workers may rest during sun or rain.
112. Other amenities like place for changing clothes, medical aid etc. can be considered wherever possible. Big local bodies at least should do this.

Agricultural Fields

113. In some parts of the country, people do not like to make use of night-soil for manuring purpose and there all the night-soil goes waste. To popularise the use of night-soil local bodies should attach agricultural fields to the dumping grounds. The manure prepared may be utilized in those fields which would serve as a model to others. It would also be paying proposition to the local bodies.

Composting

114. The local bodies should ensure proper composting in the shortest possible time.

Gram Laxmi Gas Plant

115. The use of Gram Laxmi Gas Plant may be made in villages and towns where there are no agricultural fields or where no use of night-soil manure is possible.

Cooperative Societies

116. The local authorities may consider the possibilities of establishing cooperative societies for the purchase and sale of compost.
117. If cooperative societies of scavengers come forward they should be encouraged.

Private Dumping Grounds

118. No private dumping grounds should be allowed to continue.
119. Cleaning of night-soil and refuse from the town and its proper disposal is the primary responsibility of the local bodies and should be well looked after by them.
120. So long as customary rights are not abolished the local bodies should ensure proper treatment of the night-soil at the private dumping grounds.
121. Efforts should be made to improve the conditions of work and worker at private dumping grounds.
122. If necessary, bye-laws and necessary legislation for the proper maintenance of private dumping grounds in the interest of sanitation should be passed and strictly enforced.
123. Some local bodies have already got some bye-laws etc., which in many cases, are rarely implemented. There is great necessity for enforcing the municipal bye-laws and regulations strictly.

Chapter Eight

Underground Drainage

124. At least in all the towns already having some underground drainage system, the local bodies should make serious efforts to extend the lines so as to cover the entire towns.
125. In towns which are expanding rapidly, local bodies should make serious efforts to lay the underground sewers in all new colonies side by side with the provision of other amenities.

Sewage Disposal Plants

126. Local bodies should make use of Sewage Disposal Plants. They may follow the Mudurai experiment for use and disposal of sewage of the town.
127. Local bodies can depute their officers to study the working of

the farm of Madurai and adopt similar methods with improvements, if any, in their own towns.

128. After consulting experts, growth of vegetables etc., on such farms may be encouraged.

Financial Aspects

129. A specific amount must be earmarked for the underground drainage schemes separately and at least 1/3rd of the total allocations or the National Water Supply and Drainage Schemes should be earmarked for drainage schemes.

Grant of Subsidy

130. Several towns have underground drainage schemes only on paper. Funds are not forthcoming. Such schemes are not workable without adequate subsidy from the Centre or the State Governments.

131. The Central Governments must also give liberal subsidy to the State.

132. The Ministry of Health may examine the possibility of giving subsidy under their Health and Sanitation Schemes.

133. State Governments which do not give any subsidy to their local bodies as yet must make necessary provisions for the purpose.

Connections with Underground Drainage System

134. In areas served with sewers, it should be made obligatory to connect house latrines with them. Defaulters should be dealt with properly.

Conversion of Dry-latrines Grant of Loan/Subsidy

135. There are people who due to their poor financial position are not able to convert their dry latrines into water borne ones. For such people, local bodies should give loan-subsidy to house owners according to income slabs, or in the light of local conditions.

136. The Central Government should also give aid to State Governments by giving adequate assistance.

137. Wherever underground sewers have already been laid or are

being laid, latrines must be converted into flush-out latrines and in places where flush-out latrines are not possible, latrines like septic-tanks, aqua-privies etc., may be adopted.

Rebate in Taxes

138. The local bodies may examine the possibility of allowing some rebate in house, latrine tax etc., to such of the parties who may like to convert their dry latrines into water-borne etc.
139. The rebate given at present is insufficient but should be raised so as to induce sufficient number of people to convert their latrines.

Surface Drains

140. As long as surface drains exist, arrangements for keeping them clean should continue. A set of two persons—a cleaner and a *bhishti*—may be very useful.
141. The cleaners should have proper implements with them.
142. Wherever necessary, open surface drains should be covered.
143. There should be short-phased programme to make surface drains 'pucca' and at least by the end of the Third Five Year Plan, all the surface drains must be made 'pucca'. These should be so constructed as to enable the water to flow easily and not stagnate.
144. Drains should be washed daily.
145. Sullage water from drains can be very well utilised in agricultural fields.
146. In towns where water supply is not sufficient, other arrangements for keeping the drains clean should be made.

Chapter Nine

Design of Latrines

147. People must be educated and persuaded to construct latrines in their houses. The local bodies should pass bye-laws for having latrines compulsorily and the plans for construction of new houses should not be passed unless and until there is a provision of a sanitary latrine.
148. By-laws should also be passed for compelling owners of the existing houses which are without latrines to construct latrines.

Public Latrines

149. The number of public latrines should be increased keeping in view the demand for them.
150. In places of congestion where these are needed, the umber of latrines should be increased.
151. Underground water-brone latrines at the cross roads would be found convenient.
152. As long as all the public latrines cannot be converted into water-borne ones, they should at least be made fly-proof. This would also prevent pig entry.
153. The local bodies should give serious thought to the models of public latrines. Primitive latrines must be remodeled and more sanitary latrines adopted.
154. As far as possible, public latrines should be water-borne with arrangements for the automatic flow of water after periodical intervals.
155. Public latrines should be constructed in the most crowded localities and places within easy reach. Latrines constructed at remote or far places are not likely to be of much use and whenever these are used, these are misused.
156. Wherever possible, latrines should be built on a large scale underground.

Private Latrines

157. Roof-latrines as in Punjab must be brought down. Bye-laws should be framed that in new buildings roof-latrines are not allowed. Efforts should be made through legislation etc., or persuasion to bring down existing roof-latrines. Wherever space is available, sanitary latrines be constructed; otherwise standardised receptacles should be made use of.
158. Even roof-latrines may be connected into water-borne ones and connected with underground sewers where such sewers exist or to septic-tank which could be constructed on ground floor. Where there is no space on the ground-floor or there are no sewers, septic-tanks may be constructed on co-operative basis at any available place in the street itself and three or four houses may connect their latrines with them. Local bodies can help in keeping such septic-tanks clean periodically.
159. All primitive latrines should be demolished. As long as these continue, no improvements in scavenging conditions can be

expected. For remodelling primitive latrines, not only legislation etc. but persuasion would also be required.

160. Some of the improved type of latrines suitable for towns are:

(1) Flush-out Latrine.
(2) Automatic flush-out Latrine
(3) Septic-tank Latrine.
(4) Aqua-privy Latrine.
(5) Leach type Latrine.

The local bodies should adopt the type of latrine they feel can be best suited to their towns. They may improve thereon making them more suitable to their towns according to the conditions prevailing locally.

Latrines on Market and Festival Days

161. Wherever markets or fairs are held regularly, permanent latrines should be provided which should possibly be water-borne failing which any other improved type may be adopted. Wherever receptacles are provided, their sizes and arrangements for keeping them clean must be specified. Extra scavengers should be provided on such days. Where markets etc., are held after long intervals or where permanent latrines are not possible, temporary latrines of improved types with arrangements for extra scavengers should be made.

Latrines in Rural and Semi-rural Areas

162. Latrines should be introduced in villages also progressively and people must stop present indecent and insanitary practice of easing everywhere. It is indecent to observe a large number of people squatting all along the railway lines etc. Some of the better type of latrines suitable for rural and semi-rural areas are:

(1) 'Kucha" Trench Latrine.
(2) 'Pucca' Trench Latrine.
(3) Bore Hole Latrine.
(4) Gopuri Latrine.
(5) Dhanoli Type Latrine.

Chapter Ten

Latrines of Railways

163. The recommendation made in respect of the removal of night-soil, providing of receptacles, conversion of dry latrines etc., may under a phased programme be also adopted by the Indian Railways as well.
164. Where water-borne latrines are not possible, Indian Railways could construct fly-proof latrines.
165. Indian railways have resources and technical talent to provide an object lesson to the whole country in the matter of construction of various types of sanitary latrines. This would play a great psychological role on the minds of lakhs of people who see and use them daily throughout the country.

Chapter Eleven

Customary Rights

166. Private scavengers have customary rights not only for cleaning latrines in certain localities but at some places over the possession and disposal of night-soil also. Such customary rights must not in any case be allowed to continue and be abrogated immediately.
167. The local bodies should pay reasonable compensation in cash to such of the scavengers who might be affected and do not get employment due to acquisition of customary rights.
168. The Governments at the Centre and State should come forward to finance such of the local bodies who are not in a position to pay reasonable compensation.
169. Wherever cash compensation is not possible, or the local bodies cannot offer employment under them, the affected scavengers must be taken in other services. Most of the scavengers would be prepared to forgo their customary rights if provided with municipal services.

Scavenging Cess

170. In places where there is no scavenging cess, such a cess should be levied and the scavenging services municipalised. The scavenging cess should be so adjusted that scavenging services

may become almost self-supporting.

L.S.G. Departments

171. It should be the duty of the Local Self-Government Departments to initiate necessary legislation for the acquisition of customary rights. They should handle the problem at the State level through their legislature and ensure that a uniform policy is adopted by all the local bodies and fully implemented.

Small Committees to be appointed

172. The State Governments and Union Administrations should appoint small Committees to go into the whole question of the abolition of customary rights. They should not merely study the legal aspects but go thoroughly into other problems as well connected with it. They should draw up a phased programme for the abolition of customary rights and see that the work is completed as soon as possible but not later than a fixed target date. They should determine the basis, mode of payment and the amount etc.
173. Negotiations and mutual agreements between the local bodies and scavengers can bring the desired effects. The local authorities would be well advised to tackle the problem in a friendly and persuasive manner. Co-operation of non-official organisations and social workers interested in welfare of scavengers may also be enlisted.

Chapter Thirteen

Administrative Set-up

174. The problem of improving the scavenging conditions should be dealt with by the Local Self-Government Departments of the States/Union Administrations or such other Departments as may be concerned with it.
175. All the necessary legislation, policy, directive principles etc., should be issued by the Local Self-government Departments and they should evolve a machinery to ensure the implementation of their directives etc., fully and properly through the local bodies and such of their officers as they may like to appoint. Mere enacting of legislation or issuing directives

would not be so much helpful as their proper implementation.

176. The State Governments/Union Administrations should appoint Special Officers exclusively to look after the problems of scavenging and scavengers.

Need for Central Officer

177. The Central Government must appoint an officer having the necessary background knowledge of the various problems of scavenging and scavengers in the country at the Centre to deal with these problems and watch their day to day development.

178. A Small Vigilance Committee consisting of non-officials who are well acquainted with the magnitude of the problem and have shown serious interest in tackling it, should be appointed by the Ministry of Home Affairs to move about from State to State, meet and discuss with officials, social workers and members of the public about the urgency of the problem and mobilise their efforts for solving it.

179. The Ministry of Home Affairs who are concerned with the Welfare of Scheduled Castes should initiate legislation for the abolition or customary rights, if necessary.

180. The ceiling for water supply and drainage schemes should be fixed separately. The ceiling for drainage scheme is too small in comparison to the cost and requirements of this scheme. The Ministry of Health should consider increasing this ceiling.

181. The Ministry of Health may consider giving some grants/ loans to be recovered in convenient instalments to the local bodies through State Government for the purchase of vehicles etc., for transportation of night-soil and town refuse.

182. The Ministry of Food and Agriculture (Agriculture Department) may also consider what possible help they can give for converting night-soil into manure under agricultural programmes.

183. The State Governments/Union Administrations should help their local bodies technically, financially and otherwise.

184. The Co-operation Department of the Ministry of Community Development and Co-operation can also tender their advice and start some of the schemes if possible for the welfare of scavengers particularly those who may be affected due to abolition of customary rights, and also help in the collection, preparation and sale etc., of the manure on a co-operative basis.

Chapter Fourteen

Some General Aspects

Though the question of the improvement of the living conditions and environment of scavengers is beyond the scope of the terms of reference of the Committee, yet as this is an important aspect having direct bearing on the problem, some suggestions have been made which require urgent attention.

Housing

185. Quarters for conservancy staff should be as near their places of work as possible but never around the pail-depots, morgues, public latrines, dumping grounds etc. Such location of scavenger's quarters at such places amounts to criminal neglect. A few quarters here and a few there may be built in colonies inhabited by others as well so as to avoid segregation and give an impression of a mixed colony.
186. It is of primary importance that the housing conditions of scavengers are radically improved. This would create a new sense of self-respect in them and particularly their children.
187. For the welfare of scavengers, maximum use of the Slum Clearance Schemes should be made. It is necessary to have a special provision in the plan under the Work and Housing Ministry for providing houses to the conservancy staff of the local bodies.
188. The Low Income Housing Scheme—the local bodies should construct houses of scavengers under the Low Income Housing Scheme and as far as possible pass on the quarters to them on Hire Purchase System.
189. Leaving aside the schemes under the Slum Clearance and Low Income Housing Schemes, it would be advisable for the local bodies to construct houses for scavengers and give them on Hire Purchase System. This would enable the scavengers to own their houses after some time and live in them permanently.
190. One-room tenements are not adequate. Scavengers have large families and even taking an average of five members per family, one-room tenements are too small. From the point of view of moral hygiene also every tenement must make provision for two living rooms. A kitchen, a bathroom, a

latrine—flush-out or any other sanitary type like septic tank etc. built-up cupboard; a slightly raised platform for fire place with a hood and chimney over it and possibly a hollow underneath for storing fuel and a verandah should be provided. Wherever possible, electricity must also be provided. Water supply is a necessity.

191. The quarters need not necessarily be completely 'pucca' and wherever the cost is on the high side, these may be 'semi-pucca' as well. The cost of each tenement for such quarters which are not to be owned by the scavengers on Hire Purchase System, etc. may vary from Rs. 1,200 to Rs. 2,000 on the lines of the quarters built in Madras under Slum Clearance Schemes.

192. It is very important that amenities like primary school, a welfare centre with a community hall which may be used as community centre, medical facilities etc., are provided in large colonies. A social worker should invariably be provided to be incharge of the welfare activities. There is a need for appointing a social worker for covering a population of 100 families.

Conditions of Service

193. At present, conditions of service in different States are far below the minimum needs which is the principal reason for discontentment amongst scavengers. The better the service and working conditions, better would be the relations between the scavengers and the local bodies.

Wages

194. Wages would be paid to scavengers according to the Minimum Wages Act.

195. The State Governments and Union Administrations who have not done it so far, must appoint small Committees to go into the whole question of wages and payment of other allowances to scavengers and suggest basic wages which should, however, not fall below the minimum wages. Question of payment of other allowances as may be admissible to scavengers must also be considered.

Unclean Work Allowance

196. Unclean Work Allowance should not be in cash but in the

form of extra amenities such as gloves, water etc. to make the occupation more clean. Extra allowance paid in cash is not utilised properly and the work remains unclean. Such a terminology should not be used. Unclean Work Allowance can very well be called like any other ordinary allowance.

Distribution of Wages

197. Wages should be distributed in the presence of a responsible officer of the local body who should ensure that no creditors are allowed to throng the premises and harass the scavengers. Such an officer should also keep an eye on the supervisory staff of the local bodies directly in touch with the scavengers and who are often alleged to accept illegal gratifications.

Hours of Work

198. For full time employees, the hours of work should not be more than 7 hours per day of which only 4 to 5 hours may be reserved for scavenging work. Part-time worker should not be asked to work both in the morning and evening as this deprives them of any other work and causes them hardships.

Roll Call

199. The scavengers on duty should be called only twice for roll call—once in the morning and once in the afternoon.

Leave Rules

200. Local bodies should frame leave rules and the scavengers should be given the benefit of casual, privilege, earned leave etc.
201. Women scavengers must be given prenatal and post-natal leave. They should be given leave in case of miscarriage etc., also.

Off-day in a Week

202. State Governments/Union Administrations should see that at least in local bodies where the scavenging services is municipalised, scavengers are allowed a full day off in a week.

Uniforms

203. All the local bodies should give sufficient and proper summer and winter uniforms both to male and female scavengers.
204. Proper kit for scavengers for cleaning underground drains should be supplied.

Permanency of Service

205. Every effort should be made by the local bodies to give some security of employment to the temporary scavengers and avoid keeping them on a temporary basis for an indefinite period.

Full-time and Part-time Workers

206. Local bodies, as far as possible, should not appoint part-time scavengers. Full-time worker is bound to be more efficient.

Appointment of Scavengers in other occupations

207. Scavengers who may like to leave their occupations should be appointed on other posts such as peons in other Departments, 'chowkidars', 'Naka-walas' etc.
208. Vocational training may be imparted to such of the scavengers who may like to settle in a trade. After training some subsidy/ loan may be given to them for settling in trades. They be given chances and opportunities for training as drivers, cleaners and mechanics etc., in municipal workshops and other gainful occupations.

Water Supply

209. Sufficient water for drinking and washing must be supplied to scavengers in or near *basties* at times when they return from work.

Shop

210. Arrangements for supply of soap at the place of work like pail-depots and dumping grounds etc., should be made.

Women Workers

211. Creches should be started in the colonies of scavengers where a paid worker preferably a trained lady worker may be appointed to look after the children. Children must be taught clean ways of living and habits so as to make them clean persons.

Community Centres

212. There should be established Community Centres where care of children is combined with training of women in some crafts and adult education. Such community centres should also be used for cultural purposes and amusements etc.

Education

213. State Governments and local bodies should ensure that children of scavengers are provided easy admissions in schools and hostels irrespective of their caste. Efforts should be made to open primary school near their *basties* where children of scavengers and other communities may receive education.

Occupational Diseases

214. All State Governments etc., should conduct a medical survey and find out if the scavengers suffer from any disease due to their handling night-soil. If so, proper medical arrangements for them should be made. Scavengers must at least be examined periodically.

Name for Scavengers

215. In order to create a healthy atmosphere and psychological background, the scavenger should not be called by the age old name. He should be known by some more decent terminology according to local conditions and language.

The Pig

216. Rearing of pigs in any area of towns must not be allowed if anybody wants to rear piggery he should do so on the proper

lines away from towns.

Urinals for Men and Women

217. Adequate number of urinals, preferably with automatic water arrangements, should be put at public places like bus stands, market places, parks, cinema halls, thoroughfares etc. Urinals for women must also be provided in sufficient numbers. Urinals near latrines are bound to be misused and should be avoided. Efforts should be made to equip men's urinals with Western type basins which cannot be misused.

Dust Bins

218. Local bodies must provide dust bins at all suitable places which should be within easy reach and ensure their proper cleaning. The dust bins may be kept fixed on paved floors.

Legislation

219. State Governments/Union Administrations should explore possibilities of framing rules, bye-laws etc., which they may consider necessary for the solution of various problems.

Propaganda

220. Various process of scavenging problems such as use of defective and good latrine, methods of scavenging, working at pail-depots, dumping grounds and system of carriage etc., can be filmed by the Ministry of Information and Broadcasting in consultation with the Ministry of Home Affairs and the good and bad spots in the country presented before the masses who should be educated in the use of improved methods.

221. The documentaries, cinema slides etc., prepared by the Films Division of the Ministry of Information and Broadcasting would be of great help in creating public opinion for improving the conditions of scavenging and those of scavengers.

222. The Ministry of Home Affairs and the State Governments/ Union Administrations should ensure that those non-official organisations who get financial assistance either from the Centre or the State Government must include schemes for the improvement of the conditions of scavenging and scavengers

in their programmes. They should be asked to publish posters, leaflets etc., propagating the need for improvement of the conditions. The organisations who have got cinema vans should be asked to make use of such vans for the purpose of schemes for putting an end to the practice of carrying night-soil as head loads. Conversion of dry latrines into sanitary ones, schemes of improvement of the scavenging conditions and those of scavengers should be one of the compulsory schemes seeking assistance from the centre and the State.

223. It is essential that State Governments and the Ministries of the Government at the Centre should draw up a phased programme for implementing the various recommendations and suggestions in a systematic manner within the end of the Third Five Year Plan.

ANNEXURE III.3

RECOMMENDATIONS OF COMMITTEE ON CUSTOMARY RIGHTS OF SCAVENGERS (1965)

1. At the very outset the social worker was confronted with the question whether the scavengers themselves were willing to be emancipated. The resistance arose from two factors : (a) social and psychological and (b) economic. Scavenging has been a way of life for the family. A fatalistic attitude pervades the whole outlook due to the lack of education and the absence of other openings for employment. The resistance to change is illustrated by the extremely slow progress in the abolition of carrying night-soil as head-loads. A strong movement for social reform and a series of measures of assistance and employment reservations have had some impact but the vast majority of sweepers and scavengers continue in the state of backwardness.
2. The main objection to the abolition of dry latrines has been economic, arising out of fear of loss of hereditary basis. The collection of night-soil is undertaken by municipal agencies in the Eastern States and in a large number of municipalities in some other States. In Patna, for instance, the municipal corporation employed 100 scavengers to clean 35,000 latrines, an average of 35 to 300 latrines per head per day. But in a number of States such as Punjab, Rajasthan, Delhi, Uttar Pradesh, Madhya Pradesh and parts of Gujarat, a system of customary, *jajmani,* etc. continues. That includes:

   (a) exclusive rights of certain scavenger families to clean latrines in specified households and localities, and
   (b) the right to sell, mortgage or transfer as gifts the task of disposal of night-soil and other rubbish collected by the scavengers of the respective area.

3. The Malkani Committee pointed out that the scavengers doing cleaning work in private households either on their own or under the employment of some other scavengers have come to possess these rights according to which they retain, sell or mortgage the rights in the same manner as one does with his property. The Committee recommended compensation for loss

of customary rights either in cash, or by employment in the municipal services.

4. The Committee found that where scavenging is not municipalised and latrines are cleaned privately, one particular scavenger acquires hereditary right to clean such latrines as against another scavenger, by a tacit understanding and agreement. A customary relationship also develops with the householder and the scavenger receives payments in some form or the other.

   In small towns he is generally paid in kind (a daily *roti*) and some perquisites like food or clothes, etc., on some special occasions like births, marriages, deaths, etc. varying with the status of the householder. Often in bigger towns he is also paid partially in cash by each family served by him and in cities he is generally paid only in cash. The householders are unable to change the scavengers when desired. But a scavenger may enter into a deal with another on payment of an amount according to the status of the household. There have been cases of mortgage but sales are stated to be rare. However, the Committee did not consider these rights to have the essential attitudes to valid custom, nor could the right be enforced in court of law. In fact, the High Courts of Allahabad, Madhya Pradesh have ruled against the enforcement of such a right. As regards the right to undertake the disposal of night-soil, it is only in Punjab that Section 54-B of the Municipal Act, 1911, seems to indicate that property of the task of removal of night-soil may vest in a person other than a Municipality and that the Municipality may acquire this right on payment of some compensation.

5. All the same, the Committee felt that in recognition of the services rendered by scavengers and their forefathers, some sort of compensation or relief or ex gratia grant would appear to be necessary and reasonable. The question was, however, largely seen by the Committee in the light of its proposal for municipalisations of the services and it was expected that the private scavengers would be offered employment or some compensation. But in the event of conversion of a dry latrine into a water seal one, the whole business of removal of night-soil by another scavenger or by the municipality disappears. The weak customary right becomes parctically extinct. No compensation has been paid for the conversion of latrines in a number of cities and towns in Madhya Pradesh but the social

reform activities of the Harijan Sevak Sangh have helped in the change of occupation. The problem is different, however, where the municipal agency employs a contingent of scavengers, who cannot be retrenched just because there are no latrines to be serviced.

ANNEXURE III.4

SPECIAL CAMPAIGN DURING GANDHI CENTENARY YEAR

No. F. 5-7/UCD
GOVERNMENT OF INDIA,
MINISTRY OF HEALTH, F.P. & URBAN DEVELOPMENT
(Department of Health & U.D.)

NIRMAN BHAWAN
New Delhi, the 10th June 1968

To

The Chairman/President
Municipal Council/Board

To All (With population above 20,000)

SUBJECT : Special campaign for conversion of dry latrines into flush out latrines.

Sir,

1. One of the most difficult problems that the urban local bodies have to face is the question of providing facilities for scavenging and removal and disposal of night-soil. This question has been considered at various levels from time to time but has defied a complete solution. A special Committee, popularly known as Malkani Committee, studied the matter in detail and made a number of recommendations covering all aspects of amelioration of service and living conditions of scavengers. A number of recommendations of this Committee have received attention and several steps taken to improve the working conditions of scavengers over the past few years. It is nevertheless a fact that the degrading practice of manual handling of the rubbish and night-soil still persists in all the cities and towns. The only effective solutions to this problem lies in the conversion of dry type latrines into water-borne ones. It has been that even where sewers are laid, the house connections are not taken with the result that, on the one hand, the unclean practice of handling of night-soil continues and on the other we fail to make full use of the heavy investment in the sewerage schemes. This unhappy situation

calls for immediate attention of and resolute action by the authorities concerned with a view to ensure maximum utilisation of the facilities available. In case of non-sewered areas, however, different steps are required to be taken, as indicated later in this letter.

2. This matter has assumed great importance in view of the fact that the country will be celebrating the Birth Centenary of Mahatma Gandhi on October 2, 1969. There can be no better and befitting tribute to his memory than to launch a concerted programme to achieve notable progress by way of getting existing dry type latrines converted into water-borne ones in as large numbers as possible in all the urban areas by that date. Your municipality can take immediate action as suggested below to achieve maximum success.

(a) No new building should be allowed to be constructed without provision of flush out latrines. It may be pointed out that the provision of flush out latrines in a house does not involve heavy expenditure.

(b) In areas where sewers have been laid, strict enforcement of law for compulsory connections of latrines in the existing houses to sewers should be launched.

(c) Where sewers are not laid, a special drive should be launched for conversion of dry latrines into flush out latrines in selected areas by connecting them to local septic tank or leaching pits. For the guidance of the local bodies, a small brochure indicating the designs and techniques of cheap hand flush latrines has been printed and a copy of the same is enclosed. The National Committee for Gandhi Centenary has also published some literature and circulated it to various towns, through Harijan Sevak Sangh.

3. In this connection, a copy of letter No. F-7-1/67-PHE dated 19.7.1967 indicating the pattern of assistance available for both sewered as well as non-sewered areas is enclosed. Your Municipality can immediately prepare scheme under the Urban Water Supply and Sanitation Programme and submit the same to the State Government for sanction. The State Governments have powers to sanction such schemes upto a certain limit without reference to the Government of India.

4. It is hoped that you will take personal interest in this important

matter and initiate action immediately. The Ministry will be ready to render any assistance possible.

Please acknowledge receipt of this letter and let this Ministry know the action taken or proposed to be taken as early as possible.

Yours faithfully,
Sd/-
(Deva Raj)
Officer on Special Duty

No. F-5-7/UCD

Copy to the Secretary to the Government of all States & U.Ts.. L.G.S. Deptt. ________in continuation of this Ministry's letter No. F-6/68-PHE dated 18 May, 1968.

Sd/-

Officer on Special Duty

ANNEXURE III.5

CENTRALLY SPONSORED SCHEME OF MINISTRY OF WORKS & HOUSING

NO. 0.11019/8/74-PHE
GOVERNMENT OF INDIA
MINISTRY OF WORKS AND HOUSING
New Delhi, the 22nd March, 1975

To

All State Governments/Union Territories

SUBJECT : Conversion of dry latrines into sanitary latrines during Fifth Five Year Plan

Sir,

I am directed to say that the Government of India has instituted a pilot scheme under Central Sector for conversion of dry latrines into sanitary latrines during the Fifth Five Year Plan. The details of the scheme are given in the annexure. The salient points of the scheme drawn up are given below :

1. Objectives

i. Abolition of manual handling of night-soil as head load.
ii. Demonstration of cheaper method of safe disposal of spent water and human wastes in lieu of complete underground sewerage system and treatment.
iii. Improvement of sanitation in the country.

2. Criterion

The scheme is expected to be taken up in 30 selected towns in the country with the population range of 20,000 to 50,000 having a minimum of 25 gallons of water supply per capita per day.

3. Methodology

The scheme provides sewer for conveyance of waste water and septic tanks for collection and disposal. The septic tanks will be located at convenient points serving 300 to 500 population and effluent from the septic tank will be collected in a collecting well and will be removed therefrom periodically by a mobile vacuum van for safe disposal.

4. Central Assistance

During the Fifth Five Year Plan an outlay of Rs. 4.4 crores has been made in the Central Sector for this scheme. The Government of India will provide cent per cent grant assistance to each selected town towards the work of provision of laying sewers, contraction of sewer appurtenant structures, septic tanks with the exception of the latrine construction and sewer connection. One mobile vacuum will be provided for the periodical cleaning of collecting wells.

The cost of providing sewerage, septic tanks to a population of 10,000 will be about Rs. 13 lakhs. Cost of a vacuum can well be about Rs. 75.000 the total cost for each town will come to about Rs. 13.75 lakhs or say 14 lakhs for a population of 10,000 persons, the cost works out to Rs.140 per capita.

5. State Government's Liabilities

Under this scheme the State Government will have the following liabilities:

1. The Government will provide advance to local bodies for giving loans to the needy house owners for meeting expenditure on latrines construction and sewer connection, recoverable in instalments with interest. The well-to-do class is likely to be interested to avail themselves of this facility by meeting expenditure for fitting the pan and sewer connections but the weaker section may not have resources to meet this expenditure in part even. To meet this difficulty it will be necessary that in the case of persons having income below Rs. 350 per month, the entire amount of expenditure on the latrine and sewer connection is given as loan recoverable in instalments in five years with interest. In case of persons having income more

than Rs. 350 per month the loan may be given to the needy persons as an incentive but the loan should not be more than 50 per cent of the estimated cost repayable in 2 years in instalments with interest.

ii. The State Governments will lay down a procedure to ensure that the loan given by the local body is not diverted for any other purpose.

iii. After the scheme is implemented, the responsibility of maintenance of the scheme will be of the local body. Therefore, it would be necessary that the resources of local body are augmented by raising a scavenging/sewage tax not less than 1 per cent of the rateable value of the property to meet the maintenance expenditure. The scheme should be sanctioned only after the concerned local body adopts a resolution to this effect.

iv. The State Government will have also to ensure and see that all the dry latrines in the selected areas are converted to sanitary latrines and connected to the sewer by the individual house owner.

The guidelines for the selection of the towns for implementation of the scheme are enclosed.

It is requested that the State Governments may, while submitting their schemes, should keep the above guidelines in view.

Yours faithfully,
Sd/-

(Prakash Narain)
Deputy Secretary
to the Government of India

ANNEXURE IV-1

ANNEXURE - IV.1

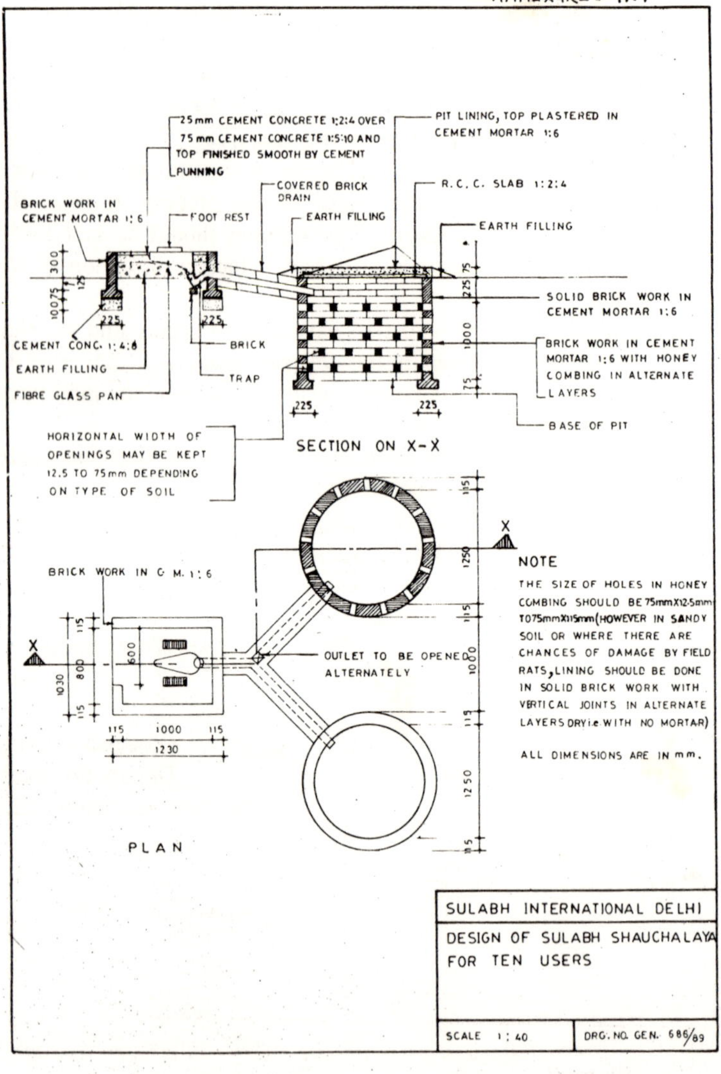

Design of Sulabh Shauchalaya (Rectangular Type)

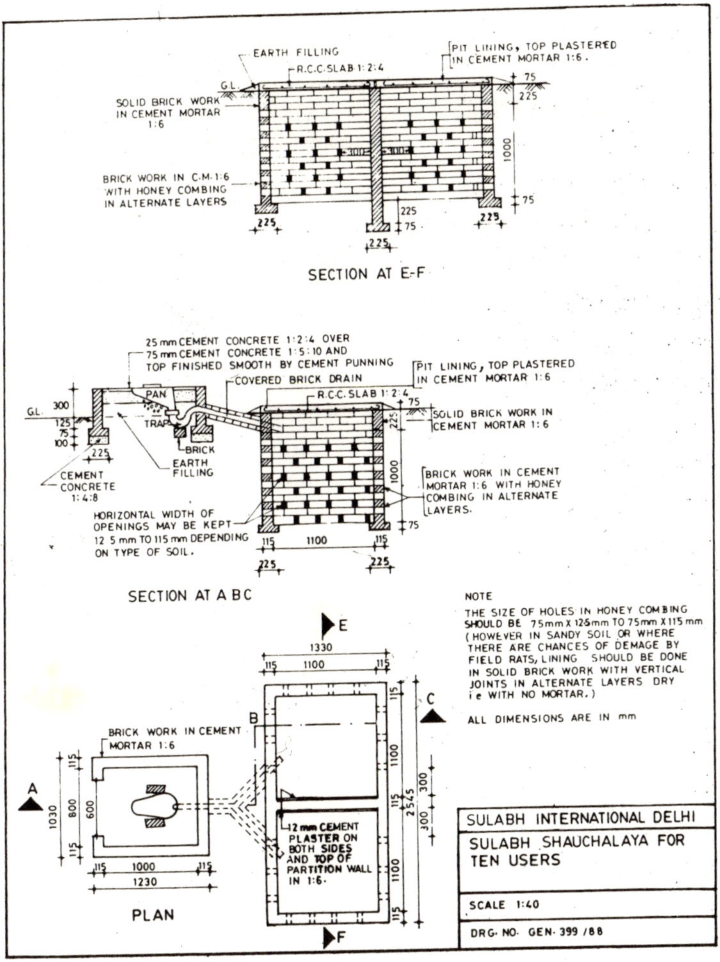

Design of Sulabh Shauchalaya (Circular Type)

ANNEXURE IV.2

NAMES OF THE TOWNS TAKEN UP UNDER THE CENTRALLY SPONSORED SCHEME OF LIBERATION OF SCAVENGERS PROGRAMME UNDER THE MINISTRY OF HOME AFFAIRS/WELFARE
From 1980-81 to 1989-90

S. No.	Name of the State	Name of the towns/municipality	Year of selection
1	2	3	4
1.	**Andhra Pradesh**		
		1. Warangal	1981-82
		2. Eluru	
		3. Rajamundry	1984-85
		4. Siddipet	
		5. Kurnool	
		6. Hyderabad	1985-86
		7. Yemiganaur	
		8. Jagtial	1986-87
		9. Bobbli	
		10. Adoni	
		11. Mandapet	1987-88
		12. Jangaon	
		13. Pithapuram	
		14. Kothagudem (NAC)	1988-89
		15. Paravathi Puram	1989-90
		16. Samalkot	
		17. Nidadavole	
		18. Korutla	
		19. Ramachandrapuram	
		20. Repalle	
		21. Srikalashsti	
		22. Ponnur	
		23. Ramagundam (NAC)	
		24. Narasapuram	
		25. Ichapuram	
		26. Amadalavalasa	
		27. Pedana	
		28. Narasaraopet	
		29. Markapuram	

ANNEXURE IV.2 *(Contd)*

1	2	3	4
	Andhra Pradesh *(Contd.)*	30. Kandukur	
		31. Punganoor	
		32. Gudur	
		33. Bellampalli	
		34. Palvancha	
		35. Mamareddy	
		36. Yellandu	
2.	**Arunachal Pradesh**		
		37. Itanagar	1987-88
		38. Pasighat	
		39. Alang	
3.	**Assam**		
		40. Nalbari	
		41. Mangaldoi	1981-82
		42. Karimganj	1983-84
		43. Hailakandi	
		44. Sapatgram	
		45. Kharupatia	1989-90
		46. Hajai	
		47. Golaghat	
4.	**Bihar**		
		48. Biharsharif	
		49. Prunea	1980-81
		50. Mudhubani	
		51. Daltanganj	1981-82
		52. Chaibasa	
		53. Bhagalpur	
		54. Gaya	1982-83
		55. Chapra	
		56. Muzaffarpur	
		57. Hazaribagh	1983-84
		58. Motihari	
		59. Bettiah	1985-86
		60. Arrah	
		61. Darbhanga	
		62. Sitamarhi	1986-87
		63. Katihar	
		64. Monghyr, Dist Monghyr	
		65. Jamui, Distt Monghyr	1989-90
		66. Samastipur Distt. Samastipur	

ANNEXURE IV.2 *(Contd)*

1	2	3	4
	Bihar *(Contd.)*	67. Giridih, Distt. Giridih	
		68. Nawadah. Distt. Nawadah	
		69. Hisua, Distt. Nawadah	
		70. Siwan, Distt. Siwan	
		71. Gopalganj, Disst. Gopalganj	
		72. Mahnar, Distt. Vaishali	
		73. Sultanganj, Distt. Bhagalpur	
		74. Narkatiaganj, Distt. W. Champaran	
		75. Chatra M. Distt. Hazaribagh	
		76. Kodarma NAC, Distt. Hazaribagh	
		77. Nokha NAC, Distt. Rohtas	
		78. Dehri Dalmia Nagar, Distt. Rohtas	
		79. Simdega NAC Gumla	
		80. Aurangabad M. Distt. Aurangabad. M.	
5.	**Haryana**		
		81. Hodal	1982-83
		82. Gharaunda	
		83. Bawal	
		84. Bawanikhera	1984-85
		85. Taraori	1987-88
		86. Pundri	
		87. Julana	
		88. Barwala	
		89. Taoru	1988-89
		90. Indri	
		91. Naraingarh	
		92. Beri, Distt. Rohtak	1989-90
		93. Buria, Distt. Ambala	
		94. Siwani, Distt. Bhiwani	
		95. Narnaund, Distt. Hissar	
		96. Kanina, Distt. Mohindergarh	
6.	**Himachal Pradesh**		
		97. Shimla	1983-84
		98. Mandi	
		99. Nahan	1984-85
		100. Chamba	
		101. Sundernagar	1986-87
		102. Kullu	
		103. Dharmasala	

ANNEXURE IV.2 *(Contd)*

1	2	3	4
	Himachal Pradesh *(Contd.)*	104. Paonta	
		105. Dalhousie	
		106. Rampur	1987-88
		107. Jwalamukhi	
		108. Dehra	
		109. Palampur	
7.	**Jammu & Kashmir**		
		110. Jammu	
		111. Udhampur	1985-86
		112. Srinagar	
		113. Reasi	
		114. Anantnag	1987-88
8.	**Karnataka**		
		115. Basavakalya	
		116. Nanjangud	
		117. Kushalnagar	1983-84
		118. Tiptur	
		119. Gadag Betanguri	
		120. Begalkote	1987-88
		121. Atnai	
		122. Rakchur	1989-90
		123. Tumkur	
		124. Shahabad	
		125. Chamarajanagar	
		126. Kippal	
		127. Navalgund	1989-90
		128. Gokak	
		129. Chickedi	
		130. Manvi	
		131. Jamkhandi	
9.	**Kerala**		
		132. Calicut	1981-82
		133. Cochin	
		134. Palghat	1989-90 nil
10.	**Madhya Pradesh**		
		135. Raipur	
		136. Shajapur	
		137. Bilaspur	1982-83
		138. Durg	
		139. Satna	

ANNEXURE IV.2 *(Contd)*

1	2	3	4
	Madhya Pradesh *(Contd.)*	140. Jabalpur	1983-84
		141. Datia	
		142. Burhanpur	1984-85
		143. Khandwa	
		144. Katni	1984-85
		145. Sehore	
		146. Rewa	
		147. Ujjain	1980-81
		148. Gwalior	
		149. Ratlam	
		150. Sagar	1987-88
		151. Narsinghagarh	
		152. Itarsi	
		153. Tikangara	
		154. Chindwara	
		155. Begumganj	
		156. Kharagaon	
		157. Siwani	
		158. Rajhadgaw	
		159. Chhatarpur	1988-89
		160. Hoshangabad	
		161. Sorngpur	
		162. Muraina	
		163. Dewas	
		164. Guna	
		165. Shivpuri	
		166. Shapurkala	
		167. Nagda	
		168. Pipperia	
		169. Damoh	1989-90
		170. Gandhi Malehra	
		171. Dindoria	
		172. Mandla	
		173. Raigarh	
		174. Kukshi	
		175. Badwani	
		176. Banda	
		177. Udaipura	1989-90
		178. Narsighpur	
		179. Dhamtari	

ANNEXURE IV.2 *(Contd)*

1	2	3	4
	Madhya Pradesh *(Contd.)*	180. Sohanapur	
		181. Kharagapur	
		182. Dharampuri	
		183. Jaora	
		184. Betul	
		185. Dhar	
		186. Mandsaur	
		187. Neemuch	
		188. Hata	
		189. Badnagar	
		190. Alirajpur	
		191. Balaghat	
		192. Gohad	
		193. Bada Malehra	
		194. Badi	
		195. Gadarawara	
		196. Seoni Malwa	
		197. Dabra	
		198. Khachrod	
		199. Barigarh	
		200. Bhatapara	
		201. Muggeli	
		202. Chandameta	
		203. Ashta	
		204. Sendhwa	
		205. Jatara	
		206. Deori	
		207. Naimpur	
		208. Sanawad	
		209. Tarana	
		210. Khurai	
		211. Bareli	
		212. Rajgarh	
		213. Maihar	
		214. Kailaras	
		215. Wara-Seoni	
		216. Mhow	
		217. Sankach	
		218. Budhar	
		219. Chanderi	

ANNEXURE IV.2 *(Contd)*

1	2	3	4
	Madhya Pradesh *(Contd.)*	220. Gormi	
		221. Nowgong	
		222. Shahdol	
		223. Parsa	
		224. Chaurai	
		225. Bamhani-Banjar	
11.	**Maharashtra**		
		226. Khamgaon	1981-82
		227. Udgir	
		228. Malkhapur	1983-84
		229. Kamptee	
		230. Nasik	1985-86
		231. Wardha	
		232. Kolhapur	1986-87
		233. Barchi	
		234. Akot	
		235. Achalpur	
		236. Washim	1988-89
		237. Nandurbar	
		238. Hingoli	
		239. Kotoli	
		240. Tumsar	
		241. Bhandara	
		242. Shivpur	
		243. Chalisgarh	
		244. Chopda	
		245. Yeala	
		246. Amalner	
		247. Arivi	
		248. Hingaghat	
		249. Ramtek	
		250. Shagaon	
		251. Yeotmal	
		252. Sawantwadi	1989-90
		253. Reha	
		254. Mahad	
		255. Sinnar	
		256. Manmad	
		257. Sangmner	
		258. Shrirampur	

ANNEXURE IV.2 *(Contd)*

1	2	3	4
	Maharashtra *(Contd.)*	259. Satara	
		260. Mahaswad	
		261. Phaltan	
		262. Gashilgaong	
		263. Jaishinghpur	
		264. Kagal	
		265. Islampur	
		266. Buldhana	
		267. Pusad	
		268. Digras	
		269. Karnanja	
		270. Wardud	
		271. Mershi	
		272. Ballapur	
		273. Warora	
		274. Desaiganj	
		275. Vaijapur	
		276. Ambed	
		277. Bhokardan	
		278. Ambejogai	
		279. Parti-Vaijanath	1989-90
		280. Dhanmabad	
		281. Degiloor	
		282. Sailu	
		283. Ratnagiri	
		284. Chiplun	
		285. Khed	
		286. Vasai	
		287. Pen	
		288. Uran	
		289. Alibag	
		290. Matheran	
		291. Trimbak	
		292. Daund	
		293. Barmati	
		294. Bahuri	
12.	**Manipur**		
		295. Bhishnupur	1987-88
		296. Nambal	
		297. Thoubal	

ANNEXURE IV.2 *(Contd)*

1	2	3	4
13.	**Orissa**		
		298. Bhubaneshwar	1982-83
		299. Cuttack	1983-84
		300. Bhadrak	1985-86
		301. Jaruguda	
		302. Puri	
		303. Khurda	
		304. Talchar	1986-87
		305. Bohampur	1987-88
		306. Sambalpur	
		307. Jatni	
		308. Phulbani	1989-90
		309. Koraput	
		310. Kendrapura	
		311. Pekalakhemundi	
		312. Titlagarh	
		313. Deongarh	
		314. Bhuban	
		315. G. Udyagiri	
		316. Bhanjanagar	
14.	**Punjab**		
		317. Kurali	1987-88
		318. Shankot	
15.	**Rajasthan**		
		319. Bhelwara	1981-82
		320. Makrana	1982-83
		321. Bharatpur	
		322. Nagaur	1984-85
		323. Pali	
		324. Sirohi	
		325. Hanumangarh	1985-86
		326. Chittorgarh	
		327. Sikkar	
		328. Tonk	
		329. Behror	1986-87
		330. Beawar	1987-88
		331. Deogarh	
		332. Bilara	
		333. Ajmer	1989-90
		334. Churu	

ANNEXURE IV.2 *(Contd)*

1	2	3	4
	Rajasthan *(Contd.)*	335. Pushkar	
		336. Jhalawar	
		337. Shri Ganganagar	1989-90
		338. Jaisalmer	
		339. Jalore	
		340. Banswara	
		341. Nathadwara	
		342. Sumerpur	
		343. Baran	
		344. Mount Abu	
		345. Parttapgarh	
		346. Bhinder	
		347. Shahpura (Bhilwara)	
		348. Rajasamand	
		349. Kushalgarh	
		350. Sagwara	
		351. Mandalgarh	
		352. Barmer	
		353. Alwar	
		354. Dholpur	
		355. Dungurpur	
		356. Bhinmal	
16.	**Tamil Nadu**		
		357. Udumelpet	1981-82
		358. Ambattur	1984-85
		359. Nagarcoil	
		360. Villupuram	
		361. Tiruchendur	
		362. Thirukkalukundram	1985-86
		363. Veerapanchatram	
		364. Harur	
		365. Tirupparamkundram	1987-88
		366. Tiruvaiyaru	
		367. Tirupattur	
		368. Katumannarkoil	1988-89
		369. Pallipalyam	
		370. Perambalaour	
		371. Komayagounderapatti	
		372. Suleeswaranpatti	
		373. Thiruthangal	

ANNEXURE IV.2 *(Contd)*

1	2	3	4
	Tamil Nadu *(Contd.)*	374. Melapalayam	1989-90
		375. Shenocottah	
		376. Melur	
		377. Thriuvathipuram	
		378. Bodinayakanur	
		379. Devakottai	
		380. Goldenrock	
		381. Nellikuppam	
		382. Ramipet	
		383. Chidambaram	
		384. Ambathur	
		385. Cumbum	
		386. Thirumangalam	
		387. Srirangam	
		388. Dharapuram	
		389. Rasigpuram	
		390. Myladuthurai	
		391. Koilpatti	
		392. Playamkottai	
		393. Tirupathur	
		394. Villupuram	
		395. Bolur	
		396. Peranamallur	
		397. Cholapuram	
		398. Vadkkuvalliur	
		399. Ayakudi	
		400. Sular	
		401. Alathur	
		402. Pennadan	
		403. Uthmapalyam	
		404. B.P. Agrahram	
17.	**Tirpura**		
		405. Agartala (Phase I)	1981-82
		406. Agartala (Phase II)	1983-84
		407. Dharamnagar	
		408. Udaipur	1984-85
		409. Kalashar	1987-88
		410. Kumarghat	
		411. Sanamura	
		412. Belonia	

ANNEXURE IV.2 *(Contd)*

1	2	3	4
	Tripura *(Contd.)*	413. Amarpur	1989-90
		414. Sabroom	
		415. Khowai	
		416. Kamalpur	
		417. Telamura	
18.	**Uttar Pradesh**		
		418. Barabanki	1981-82
		419. Badaun	1985-86
		420. Sitapur	1986-87
		421. Jalaun	
		422. Mirganj	
		423. Muzaffarnagar	
		424. Allahabad	1987-88
		425. Faizabad	
		426. Varanasi	
		427. Haridwar	
		428. Nagina	1987-88
		429. Bhadohi	
		430. Ghaziabad	
		431. Narhara (Etah)	1889-90
		432. Ballia (Ballia)	
		433. Sikandar Roa (Aligarh)	
		434. Pukhrayan (Kanpur)	
		435. Bachhraon (Muradabad)	
		436. Bairari (Muradabad)	
		437. Bijnour (Bijnour)	
		438. Haldaur (Bijnaur)	1989-90
		439. Etawah (Etawah)	
		440. Bharathinana	
		441. Kaimganj (Farrukhabad)	
		442. Jhansi (Jhansi)	
		443. Maranipur (Jhansi)	
		444. Rath (Himirpur)	
		445. Lalitpur (Lalitpur)	
		446. Fatehpur (Fatehpur)	
		447. Changar (Mirzapur)	
		448. Basti (Basti)	
		449. Khalibabad (Basti)	
		450. Bauthnwa (Gorakhpur)	
		451. Deoria (Deoria)	

ANNEXURE IV.2 *(Contd)*

1	2	3	4
	Uttar Pradesh *(Contd.)*	452. Gaurabarhat (Deoria)	
		453. Jais (Raebareli)	
		454. Gangaghat (Unnao)	
		455. Nawabganj (Gonda)	
		456. Gonda (Gonda)	
		457. Sultanpur	
		458. Pratapgarh (Pratapgarh)	
		459. Shikapur	
		460. Bulandshar (Bulandshahar)	
		461. Pauri (Pauri)	
		462. Rudrapur (Nainital)	
		463. Almora (Almora)	
		464. Bageshwar (Almora)	
		465. Balsi (Badaun)	
		466. Bisauli (Badaun)	
		467. Ramnagar (Varanasi)	
		468. Neemsar Mishrik (Sitapur)	
		469. Akbarpur (Faizabad)	
		470. Garhmukteshwar (Ghaziabad)	
19.	**West Bengal**		
		471. Sonamukhi	1982-83
		472. Murshidabad	
		473. Shantipur	1983-84
		474. Ghatal	
		475. Bolpur	
		476. Rampurhat	1985-86
		477. Mekliganj	
		478. Alipurduar	
		479. Burdwan	1986-87
		480. Asansol	
		481. Taki	
		482. Mathbhanga	
		483. English Bazar	1987-88
		484. Jalpaiguri	
		485. Kalna	
		486. Chakda	1988-89
		487. Jharagram	
		488. Tamluk	
		489. Durajpur	
		490. Dhuliah	

ANNEXURE IV.3

PERCENTAGE OF HOUSEHOLDS HAVING TOILET FACILITIES IN 1981 AND 1991

India/State/U.T.	Urban		Rural	Total
	1981	1991	1991	1991
1	2	3	4	5
India	58.15*	63.85**	9.48*	23.70**
India @	57.44	63.58	8.84	23.55
State				
Andhra Pradesh	44.07	54.60	6.62	18.40
Arunachal Pradesh	64.56	75.05	42.62	47.42
Assam	N.A	86.06	30.53	37.43
Bihar	52.95	56.54	4.96	11.75
Goa	49.51	55.82	29.99	40.65
Gujarat	60.11	65.71	11.16	30.69
Haryana	58.09	64.25	6.53	22.45
Himachal Pradesh	55.12	59.98	6.42	12.39
Jammu & Kashmir	64.54	N.A	N.A	N.A
Karnataka	53.28	62.52	6.85	24.13
Kerala	59.14	72.66	44.07	51.28
Madhya Pradesh	52.73	53.00	3.64	15.07
Maharashtra	59.37	64.45	6.64	29.56
Manipur	62.69	70.16	33.02	43.13
Meghalaya	70.15	85.69	18.13	31.11
Mizoram	24.52	84.44	58.37	70.73
Nagaland	65.25	75.10	26.86	37.49
Orissa	41.88	49.27	3.58	9.81
Punjab	64.75	73.23	15.79	33.18
Rajasthan	56.48	62.27	6.65	19.57
Sikkim	53.15	77.69	30.30	34.97
Tamil Nadu	51.27	57.47	7.17	23.13
Tripura	95.67	96.32	62.43	67.93
Uttar Pradesh	62.06	66.54	6.44	18.02
West Bengal	77.74	78.75	12.31	31.51
Union Territory				
Andaman & Nicobar Islands	70.90	65.72	26.32	36.88

ANNEXURE IV.3 *(Contd.)*

1	2	3	4	5
Chandigarh	78.53	79.77	3.05	70.80
Dadra & Nagar Haveli	42.83	65.14	10.59	15.42
Daman & Diu	42.93	45.75	8.39	26.57
Delhi	68.02	66.64	29.60	63.38
Lakshadweep	31.62	64.65	78.88	71.02
Pondicherry	41.54	50.02	11.85	34.88

@ Excludes Assam and Jummu & Kashmir

ANNEXURE IV.4

PERCENTAGE OF SCHE[illegible] CASTES, SCHEDULED TRIBES AND OTHER HOUSEHOLDS HAVING TOILET FACILITIES - 1991

India/State/U.T.	Total			Rural			Urban		
	Scheduled Castes	Scheduled Tribes	Others	Scheduled Castes	Scheduled Tribes	Others	Scheduled Castes	Scheduled Tribes	Others
1	2	3	4	5	6	7	8	9	10
India*	11.16	5.15	38.28	7.22	4.10	40.68	28.63	11.52	68.38
State									
Andhra Pradesh	6.66	2.95	28.17	3.50	1.49	24.32	22.00	8.02	58.24
Arunachal Pradesh	51.12	47.59	62.33	43.44	41.31	73.94	53.29	44.91	76.01
Assam	33.95	27.16	78.09	12.54	9.83	73.26	41.88	34.66	87.93
Bihar	4.99	2.19	33.45	3.33	0.97	32.39	14.16	6.08	61.00
Goa	20.87	13.06	31.26	25.00	19.23	27.14	41.12	30.37	56.44
Gujarat	23.14	7.54	50.08	5.42	2.60	32.08	36.39	14.09	69.03
Haryana	7.29	3.87	23.70	-	-	-	26.30	7.34	70.13
Himachal Pradesh	5.96	3.83	35.95	19.71	18.12	54.10	14.25	6.66	64.41
Karnataka	9.24	2.59	34.78	8.90	2.71	40.53	27.66	8.06	86.30
Kerala	29.33	26.25	44.28	13.85	11.65	52.03	53.90	46.57	74.51
Madhya Pradesh	6.27	1.91	23.40	2.21	1.14	21.25	22.26	5.45	60.09
Maharashtra	16.81	4.40	43.62	7.01	2.25	39.12	33.93	7.80	67.46

(Contd.)

Annexure IV.4 *(Contd.)*

1	2	3	4	5	6	7	8	9	10
Manipur	40.43	32.49	57.47	22.95	17.36	75.61	56.39	47.76	70.46
Meghalaya	46.79	33.96	71.95	25.20	16.26	83.04	66.17	36.97	91.13
Mizoram	81.25	56.11	90.48	70.33	58.42	84.21	82.36	49.12	85.58
Nagaland	-	-	-	34.06	25.24	73.28	69.15	54.58	80.62
Orissa	3.82	1.86	20.73	1.60	0.77	15.81	15.22	5.57	58.83
Punjab	16.04	8.31	44.37	-	-	-	40.01	19.33	80.53
Rajasthan	10.62	5.11	34.04	3.10	1.00	36.35	25.39	8.57	69.12
Sikkim	26.82	23.28	55.95	35.17	30.71	81.64	35.51	30.54	78.32
Tamil Nadu	9.84	5.10	28.96	11.37	4.02	45.89	26.76	7.90	61.76
Tripura	75.26	72.09	92.31	37.24	36.02	94.42	83.82	79.37	97.27
Uttar Pradesh	6.87	2.60	38.50	24.23	7.41	59.22	21.77	7.90	71.83
West Bengal	16.61	8.44	56.75	7.83	4.18	51.63	39.36	15.27	83.33
Union Territory									
Andaman & Nicobar Islands	-	-	-	12.78	9.99	58.13	38.82	28.05	65.85
Chandigarh	47.88	1.61	55.03	-	-	-	75.13	3.37	84.35
Dadra & Nagar Haveli	29.23	22.45	72.94	3.26	2.72	19.24	48.07	37.32	82.06
Daman & Diu	13.90	7.33	24.91	3.62	1.83	11.47	30.09	9.90	47.96
Delhi	31.15	14.36	33.01	-	-	-	71.85	34.32	75.34
Lakshadweep	-	-	-	67.71	78.27	57.93	94.55	88.73	96.11
Pondicherry	13.10	6.93	26.39	-	-	-	39.89	14.03	53.74

* Excludes Jammu & Kashmir

ANNEXURE IV.5

STATEWISE ESTIMATES OF THE NUMBER OF DRY LATRINES IN 1989

State/U.T.	1981					
	Population ('000)		No. of households		Average households	
	Urban	Rural	Urban	Rural	Urban	Rural
1	2	3	4	5	6	7
Andhra Pradesh	124876	410621	2457559	8427292	5.0	4.9
Arunachal Pradesh	414	5904	9759	115188	4.2	5.1
Assam	-	-	-	-	-	-
Bihar	87190	611957	1432934	10148311	6.2	6.1
Goa	3518	7349	69076	141336	5.1	5.2
Gujarat	106017	234841	1895057	4039093	5.6	5.9
Haryana	28274	100952	509104	1486489	5.6	6.7
Himachal Pradesh	3260	39548	77693	706101	4.2	5.6
Jammu & Kashmir	12604	47270	197865	742000	6.4	6.4
Karnataka	107296	264061	1845906	4556201	5.8	5.8
Kerala	47713	206824	791933	3631344	6.0	5.7
Madhya Pradesh	105865	415924	1923102	7398024	5.5	5.6
Maharashtra	219936	407906	4156618	7534337	5.3	5.4
Manipur	3755	10455	59621	171502	6.4	6.1
Meghalaya	2413	10945	46317	209618	5.2	5.2

(Contd.)

Annexure IV.5 *(Contd.)*

1	2	3	4	5	6	7
Mizoram	1218	3719	20548	60833	5.9	6.1
Nagaland	1202	6547	24481	124999	5.0	5.2
Orissa	31103	232600	620468	4384711	5.0	5.3
Punjab	46478	121412	812939	1935514	5.7	6.3
Rajasthan	72105	270514	1257281	4567680	5.7	5.9
Sikkim	511	2653	10666	49191	5.1	5.4
Tamil Nadu	159519	324562	3239322	7041680	4.9	4.6
Tripura	2256	18275	43180	336697	5.3	5.4
Uttar Pradesh	198991	909629	3378068	15840819	5.9	5.7
West Bengal	144467	401339	2716794	7032388	5.3	5.7
Andaman & Nicobar Islands	496	1391	11368	29303	4.4	4.8
Chandigarh	4228	288	94713	5883	4.5	4.9
Dadra & Nagar Haveli	69	968	1305	18123	5.3	5.3
Delhi	57682	4522	1139862	71922	5.1	5.2
Lakshadweep	186	216	3026	3611	6.2	6.0
Pondicherry	3160	2884	59384	56406	5.3	5.1
All India	**1576802**	**5076076**	**28905949**	**90866596**	**5.5**	**5.6**

Source: 1. From Col. 2 to Col. 5; Census of India 1981 Part II.
2. From Col. 8 to Col. 9; Registrar General.
3. January to December 1983.

Source: Report of the Task Force on the Problems of Scavengers constituted by the Planning Com[illegible]

ANNEXURE IV-5 contd.

1989 Population ('000)		Estimated No. of Households ('000)		Jan.-Dec. 83 (38th round NSSO) % of households using service latrines		Estimated No. of Households using service latrines in 1989	
Urban	Rural	Urban	Rural	Urban	Rural	Urban	Rural
(8)	(9)	(10)	(11)	(12)	(13)	(14)	(15)
167167	453715	33434	92595	12.45	1.81	416253	167597
27608	211434	5020	37756	36.30	10.63	182226	401346
123715	705538	19954	115662	13.39	1.67	267184	193165
134036	263298	23935	44401	8.15	1.79	195070	79478
41760	116561	7457	17397	34.07	0.75	254060	13048
4022	45441	958	8114	24.74	0.29	23701	2353
16600	54826	2594	8567	65.36	7.53	169544	64510
145745	292014	25128	50347	5.55	0.76	139460	38264
61860	230413	10310	40423	2.59	1.37	26703	55380
149913	468439	27257	83650	22.45	1.20	611920	100380
280772	450764	52976	83475	13.97	1.39	740075	116030
5097	12153	796	1992	31.83	11.62	25337	23147
3652	12883	702	2477	37.15	-	26079	-
2241	8349	448	1605	51.38	-	23018	-

(Contd.)

ANNEXURE IV.5 (*Contd.*)

(8)	(9)	(10)	(11)	(12)	(13)	(14)	(15)
46903	257163	9381	48521	12.31	0.60	115480	29113
61151	132197	10728	29084	28.18	1.26	302315	26440
103487	320270	18156	54283	18.89	0.90	342967	48855
928	3287	181	609	10.20	-	1846	-
193313	356046	39452	77401	15.55	1.03	613479	79723
2842	21955	536	4066	-	-	-	-
289835	1019593	49125	178876	39.31	4.49	1931104	803153
178986	457873	33771	80329	16.51	2.91	557559	233757
811	1906	184	397	26.10	3.24	4802	1286
696	7174	166	1407	-	-	-	-
6559	317	1458	65	3.70	-	5395	-
85	1187	16	224	-	3.84	-	860
81244	4509	15930	716	14.07	2.17	224135	1554
4766	8302	934	1597	1.12	-	1046	-
210	244	34	41	-	-	-	-
2123	4538	360	744	3.73	1.11	1343	826
4251	2954	802	579	4.47	-	3585	-
2142380	5925343	387163	1021544	18.08	2.29	7205686	2480256

ANNEXURE IV.6

NATIONAL SCHEME OF LIBERATION AND REHABILITATION OF SCAVENGERS AND THEIR DEPENDENTS

List of Trades

AGRICULTURE AND ALLIED SECTOR

1. Bullocks and carts
2. Coconut plantation
3. Coffee plantation
4. Supply of Diesel Pumpsets and pipeline
5. Mango plantations
6. Milch cattle
7. Piggery
8. Sericulture
9. Cashew processing
10. Manufacture of mats etc.

SMALL SCALE INDUSTRIES SECTOR

1. Flour Mills
2. Silk power looms
3. Automobile workshops
4. Printing press/Book binding
5. Furniture making
6. Wooden cart making
7. Textile products
8. Table moulded bricks
9. Coir products
10. Power looms
11. Handloom weaving
12. Silk reeling
13. Stone works
14. Tobacco bran
15. Toys and doll making
16. Screen printing

SERVICE SECTOR

1. Photo Studio
2. Repair of radio/TV etc.

3. Repair of Refrigeration/coolers etc.
4. Servicing of two/three wheelers
5. Auto rickshaw
6. Cycle repair shops
7. Pumpset repairs
8. Carpentry
9. Plumbering and sanitary
10. Retail dealership
11. Blacksmithy
12. Sheet metal work
13. Tailoring
14. Pan and cigarette shop
15. Juice vendors
16. Fruit and Vegetable vending
17. Ice cream vendors
18. Newspaper and magazine shop
19. Repair of household domestic appliances like pressing iron, toaster, gysers etc.
20. Packing
21. Repair of diesel engines, electric motors etc.
22. Watch repair
23. Painting of walls, doors and windows
24. Masonry
25. Tile laying
26. Glass pane fixing
27. Shorthand and typing
28. Embroidery and needle work
29. Tyre vulcanising unit
30. Loudspeaker, amplifier, mike etc.
31. Battery charging
32. Transformer manufacturing/repair
33. Cable manufacturing

BUSINESS SECTOR

1. Electric store
2. Provision store
3. Tiles and building material shop
4. Tea shops
5. Fish business
6. Fruit/vegetable shop
7. Machine shop (grinding etc.)

8. Wood making
9. Furniture making
10. Food and vegetable preservation
11. Small transformers and coil winding
12. Welding
13. Lens grinding and polishing
14. Ceramic/pottery
15. Decoration and glass-wares
16. Cane work and bamboo work
17. Saree Printing
18. Utensil fabrication
19. Tempos, buses, trucks
20. Transport-related activities
21. Weaving
22. Garment making
23. Steel furniture
24. Washing powder making
25. Cabinet making
26. Plastic mould maker

ANNEXURE IV.7

COMPOSITION OF THE STATE-LEVEL MONITORING COMMITTEE

The composition of the State-level Monitoring Committee would be as follows:

1.	Chief Secretary	Chairperson
2.	Secretary, Department of Urban Development/Local Bodies	Member
3.	Secretary, Industries/Small Scale	Member
4.	Secretary, Industries Rural Development	Member
5.	Secretary, Labour	Member
6.	Secretary, Education	Member
7.	Secretary, Technical Education	Member
8.	Secretary, Institutional Finance	Member
9.	Secretary, Planning	Member
10.	Secretary, Finance	Member
11.	Two well-known non-officials committed to the cause of scavengers, preferably belonging to them to be nominated	Member
12.	Two senior-most available representatives of lead banks operating in the State, if necessary, by rotation	Member
13.	Secretary, Scheduled Castes Welfare	Member-Secretary

ANNEXURE IV.8

COMPOSITION OF THE DISTRICT-LEVEL MONITORING COMMITTEE

The composition of the District-level Monitoring Committee would be as follows:

1.	District Magistrate/Collector/ Deputy Commissioner	Chairperson
2.	Chief Development Officer/	Member
	District Development Officer	Member
3.	General Manager, Dist. Industries Centre	Member
4.	Poject Officer/Area/General Manager of Nehru Rozgar Yojna	Member
5.	District Village & Industry Officer of KVIC Board	Member
6.	Principals of the ITIs	Member
7.	Principals of the Community Polytechnics	Member
8.	District Welfare Officer incharge of SCs welfare	Member
9.	Assistant Registrar, Cooperative Societies	Member
10.	Representative of lead banks	Member
11.	Chief Executive Officer of local bodies	Member
12.	Two non-officials committed to the cause of scavengers, preferably belonging to them	Member
13.	District Manager/Assistant Manager/ ADDO of the State Scheduled Castes Dev. & Finance Corporations	Member-Secretary

ANNEXURE IV.9

PROGRESS MADE UNDER THE NATIONAL SCHEME OF LIBERATION AND REHABILITATION OF SCAVENGERS AND THEIR DEPENDENTS

S. No	State/UTs	No. of scavengers Identified	1992-93		1993-94		1994-95	
			Trained	Rehbtd.	Trained*	Rehbtd.	Trng.	Rehbtd.
1.	Andhra Pradesh	7473	-	989	500	2556	-	3000
2.	Assam	5651	-	-	-	61	-	-
3.	Bihar	40249	-	-	-	-	-	-
4.	Gujarat	40000	-	623	-	973	-	-
5.	Haryana	18188	-	-	888	1295	-	-
6.	Himachal Pradesh	4760	115	10	151	123	127	119
7.	J & K	3548	-	-	275	-	-	-
8.	Karnataka	5825	-	-	166	107	-	-
9.	Kerala	1339	-	-	144	116	12	94
10.	Madhya Pradesh	80072	11395	10194	19892	12383	8717	15661

(Contd.)

ANNEXURE IV.9 *(Contd.)*

1	2	3	4	5	6	7	8	9
11.	Maharashtra	126691	1248	1675	644	2792	2098	3481
12.	Orissa	17122	209	26	1080	805	905	828
13.	Punjab	26819	130	-	4102	1138	291	1513
14.	Rajasthan	87938	528	178	2147	1543	-	-
15.	Tamil Nadu	35561	-	-	-	2556	526	527
16.	Uttar Pradesh	226189	1868	4620	1065	15213	3505	29665
17.	West Bengal	30000	-	-	103	-	-	-
18.	Delhi	17420	-	505	13	-	140	196
19.	Pondicherry	476	-	93	-	93	-	-
20.	Nagaland	1800	-	-	-	-	-	-
	Total	**7,36,114**	**15,579**	**18,913**	**22,115**	**41,950**	**42,332**	**1,43,946**

INDEX